The Basics of Molecular Biology

Alexander Vologodskii

The Basics of Molecular Biology

 Springer

Alexander Vologodskii
New York University
New York, NY, USA

ISBN 978-3-031-19403-0   ISBN 978-3-031-19404-7   (eBook)
https://doi.org/10.1007/978-3-031-19404-7

This Springer imprint is published by the registered company Springer Nature Switzerland AG
The registered company address is: Gewerbestrasse 11, 6330 Cham, Switzerland

Preface

This book represents an attempt to describe the major features of life in a very short volume. It is addressed to people who have an education in areas different from biology and who want to receive some ideas about the world of molecular and cell biology. The book assumes a certain knowledge of physics, chemistry, and mathematics that corresponds to the level of high school. The primary attention is paid here to the most fundamental features of life. The book also briefly considers the developing areas of biology, which are especially important for medicine and the future of humankind.

Recommended Literature

There are many excellent textbooks on molecular and cell biology which describe the subject in detail. Three of these popular titles are shown below. Two other references are covering new fast-growing topics related to human genetics.

1. Alberts B, Heald R, Johnson A et al. (2022). Molecular biology of the cell. 7th ed. New York, NY: Norton & Company.
2. Watson J, Baker T, Bell S et al. (2013). Molecular biology of the gene. 7th ed. Menlo Park, CA: Benjamin Cummings.
3. Zlatanova J & van Holde K (2015). Molecular Biology: Structure and Dynamics of Genomes and Proteomes. New York, NY: Garland Science.
4. Reich D (2018). Who we are and how we got here. Ancient DNA and the new science of the human past. New York, NY: Pantheon Books.
5. Plomin R. (2019). Blueprint: How DNA Makes Us Who We Are. Cambridge, MA: The MIT Press.

New York, NY, USA   Alexander Vologodskii

Introduction

We do not know how life appeared on our planet. However, we do know a lot about the organization of this life, and the goal of this book is to give a brief outline of its major features. Among natural sciences, biology stays apart from physics and chemistry in one very important respect. We can deduce, at least qualitatively, properties of physical objects from the small number of the basic laws of physics. We can predict the motion of planets around the Sun, the magnetic field around electric current, the way light passes through the lens, and so on. Although it is technically difficult to deduce the properties of atoms and molecules from the laws of quantum mechanics, we are confident that this is possible in principle. Biology is different. There is no way to deduce the way of life from general principles. There is an enormous amount of specific complex problems that have to be solved by living organisms, and these problems have many different solutions. Although each of these solutions has to follow the laws of physics and chemistry, it is impractical to search for all possible solutions to a particular problem in the biological world. Life could happen in many different ways. We, however, want to know the way of life on our planet, the one very special choice of many possibilities. To learn about this way, we have to study life experimentally. It makes the task more difficult. However, the evolution of life always had certain logic behind it, and finding this logic greatly helps in the study. We will try to articulate this logic through this book.

Although we cannot deduce much knowledge about life from the general laws of physics and chemistry, we are confident that life follows these laws. This belief will be at the base of the book. Analyzing basic processes of life, we will try to consider them as physical phenomena, wherever such consideration is useful.

Complexity is not the only striking feature of life. It is also incredibly diverse. Indeed, trees do not resemble animals, and animals seem to have little in common with bacteria. We know today, however, that the most fundamental features of all living organisms are nearly identical. All living organisms consist of cells. Some of them, like animals, have many trillions of cells in their bodies, while others consist of a single cell. But in all cases, the cell represents a unit of life. It has all the necessary elements for self-reproduction, and new cells can be only formed by the division of the existing cells. All basic elements involved in cell reproduction are essentially identical in all organisms living on our planet. Through all living organisms, hereditary information is kept in DNA molecules (molecules of deoxyribonucleic acid), and this information is coded in the same way! The same intracellular mechanisms are used to read this information and to use it for cell growth and reproduction. And all cells use the same mechanism for another major process, duplication of the hereditary information before the cell division. All cells are fundamentally similar inside, regardless of the striking difference in their appearances. These fundamental features of life clearly manifest that all living cells originate from a single colony that somehow emerged on Earth more than 3 billion years ago.

Acknowledgments

I would like to thank my friends Anshel Gorokhovsky and Igor Kulikov, and my sister Elena Vologodskaia for their remarks and suggestions which helped me greatly in preparing this book.

Contents

Chapter 1
The Major Processes in the Biological World

1.1 Cells as Basic Units of Life

It was in the middle of the nineteenth century when scientists concluded that cells represent fundamental units of life, although the term "cell" appeared much earlier. The major experimental basis of the theory was observations in the optical microscope. The size of the cells is in the range of 1–100 μm, so the limited resolution of the microscopes did not allow us to study the details of the cell organization. Therefore, it seems amazing that in 1839, Matthias Schleiden and Theodor Schwann formulated the cell theory. The theory stated that all living organisms are composed of one or more cells. They suggested that each cell contains all the needed hereditary information to produce new cells. In 1855, Rudolf Virchow added one more fundamental point there. He stated that new cells can be formed only by the division of preexisting cells. Today we know that these major statements of the cell theory are absolutely correct.

Of course, animals consist of many very different types of cells, and all of them are different from single-cell bacteria. But one of the striking features of life, which Schleiden and Schwann could not suggest, is that the most fundamental processes inside all living cells follow the same universal mechanisms. In all cells, hereditary information is stored in DNA molecules that have an identical double-stranded structure in all cells. The same principle and mechanism are used for the duplication of DNA molecules during cell division. The genetic information stored in DNA is used to build proteins, and the same universal complex apparatus do it in all living cells. And the same code for recoding protein structures is used in all living organisms! So, life on our planet proceeds fundamentally in a single universal way, in remarkable agreement with Darwin's theory of evolution. Thus, if we want to understand the basic principles of life, we should, first of all, consider an organism consisting of a single cell. This universality of the way of life makes studying it much easier, of course. However, the organization and functioning of even the

© The Author(s), under exclusive license to Springer Nature Switzerland AG 2023
A. Vologodskii, *The Basics of Molecular Biology*,
https://doi.org/10.1007/978-3-031-19404-7_1

simplest cell are incredibly complex, and it remains to be a mystery how the first cell appeared on Earth.

In this chapter, we outline the major universal biochemical processes of life, storing and duplication of genetic information, and its use for the synthesis of proteins, the key molecules of the cells. Our emphasis here will be on the most general principles of these processes. More detailed consideration of various issues will be given in the following chapters. It is essential that at the level of detail used in this chapter, the description is equally applied to all living cells on Earth.

1.2 Important Facts from Chemistry and Physics

There are features of great importance for the properties of biological macromolecules, which we will consider in this chapter. Although these are not features specific only to biological macromolecules, they do not receive enough attention in traditional chemistry courses.

1.2.1 The Flexibility of Polymer Chains

All molecules consist of atoms that are bound by chemical bonds. Chemistry considers, first of all, covalent bonds between atoms. These bonds are very strong and stable, so they are not destroyed by collisions with other molecules due to thermal motion at normal temperatures. A covalent bond between two atoms originates from sharing the atoms' electrons. If one pair of electrons, one from each atom, is shared, it creates a single covalent bond. If two pairs of electrons, two from each atom, are shared, this creates a double covalent bond. It is common, when drawing the structural formula of a molecule, to show single covalent bonds by single lines connecting the atoms. The double bonds are shown by double lines, correspondingly (Fig. 1.1). The covalent bonds are very strong, and the energy of a single bond is in the range of 50–120 kcal/mol.

The angles between adjacent bonds depend on the chemical structure of the molecule and do not change much due to thermal motion. However, larger molecules can change their shape, or *conformation*, due to the rotation of their parts around

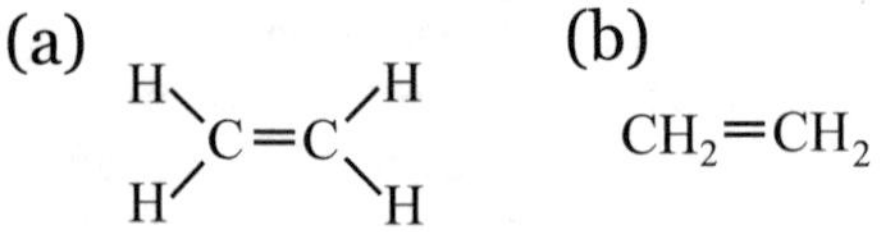

Fig. 1.1 The structural formula of ethylene. (**a**) The hydrogen atoms are connected to the carbon by the single covalent bonds, and the carbons are connected by the double bond. (**b**) A reduced drawing of the formula. Since hydrogen has a single electron, it can form only single bonds with other atoms, so the reduced drawing does not introduce any ambiguity

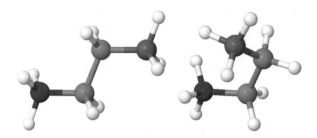

Fig. 1.2 Two different conformations of butane. Large spheres correspond to carbon atoms while small white spheres correspond to hydrogens. The conformations are obtained by rotation around the bond between two central atoms of carbon. The rotation does not change the angles between adjacent single bonds that connect carbon atoms. The image was obtained with the program ChemTube3D by N. Greeves

single bonds of the backbone. Such rotation around single bonds is restricted only

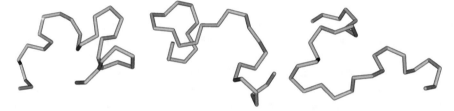

Fig. 1.3   Conformations of polyethylene $CH_3 - (CH_2)_{29} - CH_3$. Three randomly picked conformations of the molecule's backbone are shown. The carbon atoms are located at the chain vertices. The polymer molecule can adopt a huge number of possible conformations due to rotation around C–C bonds. C–H bonds are not shown here

by possible collisions of spatially close groups of atoms. For example, butane, whose structural formula is CH_3–CH_2–CH_2–CH_3, has distinguished conformations specified by the mutual arrangements of four carbon atoms. Two of these conformations are shown in Fig. 1.2.

Rotation around the single bonds creates a huge amount of possible conformations of polymer chains. This is illustrated in Fig. 1.3, where three randomly chosen conformations of polyethylene, a polymer with a repeating motive $-CH_2-$, are shown. This flexibility of polymers due to rotation around single bonds in the backbone is extremely important for the properties of nucleic acids and proteins.

1.2.2   Noncovalent Interactions Between Atoms

There are a few types of noncovalent bonds that appear in biological molecules in water solutions. These bonds are weak. Thermal motion causes permanent collisions of molecules in solution, and such collisions can easily destroy noncovalent bonds. However, if a few such bonds are formed between two groups of atoms, the probability of their simultaneous destruction becomes low. Thus, weak noncovalent

Fig. 1.4 Diagram of a hydrogen bond. The hydrogen atom forms a covalent bond with atom A_1 and a hydrogen bond with atom A_2. Atoms A_1 and A_2 can be either nitrogen or oxygen atoms

bonds are capable of stabilizing certain conformations (structures) of the macromolecules and complexes of those molecules. Noncovalent bonds are extremely important in the functioning of biological macromolecules. They can be divided into four groups.

Ionic Bonds   These bonds correspond to the electrostatic attraction between oppositely charged atoms. Although the strength of ionic bonds is greatly reduced in water, they are the strongest noncovalent bonds. Still, their strength is ten times lower than the strength of covalent bonds. The length of these bonds varies but exceeds 0.25 nm.[1]

Hydrogen Bonds   Hydrogen bonds are created by the partial sharing of a hydrogen atom between two electronegative atoms (nitrogens and oxygens). The bond strength has the largest value when it is directed along the covalent bond between the hydrogen and another atom, as shown in Fig. 1.4. The length of the hydrogen bond is close to 0.2 nm. In water solutions, the hydrogen bonds are about three times weaker than ionic bonds.

Van der Waals Bonds   This bond can appear between any two uncharged atoms which are not bound by a covalent bond. The atoms repel one another if the distance between them is shorter than a certain limit. We can say that this critical distance, r_0, corresponds to the sum of the radiuses of the atoms. The value of r_0 is usually close to 0.35 nm. It turns out that if the distance between the atoms exceeds r_0 (but smaller than $2r_0$), they attract each other, creating a van der Waals bond. Such bonds are a hundred times weaker than covalent bonds. Still, if many van der Waals bonds are formed simultaneously between atoms of a large polymer molecule (or two molecules), they can substantially stabilize a particular conformation of the molecule or a complex of two molecules.

Hydrophobic Interaction   Water molecules are highly polar, and they form a dense network of hydrogen bonds. These bonds greatly reduce the total energy of the water solution. The interfaces between water molecules and nonpolar groups of other molecules destruct the network and therefore increase the solution energy. If nonpolar groups contact one another rather than water molecules, the distraction will be smaller, and the energy of the solution will be lower. Therefore, contacts between nonpolar groups are stabilized in water solutions. Thus, there are no hydro-

---

[1] 1 nm equals 10^{-9} m. It is a common unit of length in molecular scale.

Fig. 1.5 Weak noncovalent interactions can stabilize some conformations of flexible chain molecules. One randomly picked conformation of the chain, which is not stabilized by weak bonds, is shown on the left. A compact conformation of the same chain, shown on the right, is stabilized by eight weak interactions (shown by yellow or by light gray in some formats)

phobic bonds, but there is an attraction between nonpolar molecules in water solutions due to so-called hydrophobic interaction.

Nearly all biological macromolecules are linear chemical chains. Due to the backbone flexibility, they can adopt many different conformations. Among them, there can be conformations that are well stabilized by noncovalent bonds between segments of the macromolecules (Fig. 1.5). Such conformations are critically important for cell life. During their functioning inside the cell, the majority of the macromolecules must change their conformations. These changes are only possible because the conformations of macromolecules and their complexes are stabilized by weak noncovalent bonds. This conformational flexibility would not be possible if the three-dimensional (3D) structures of the molecules are stabilized by covalent chemical bonds rather than by weak interactions. Without this conformational flexibility of the macromolecules, life would not be possible. We will return to these issues and their consequences again and again in this book.

1.3   Structure of DNA and Inheritance of Genetic Information

1.3.1   The General Principles

It is generally accepted that modern biology started in 1953 when Watson and Crick suggested the structure of DNA. Their model was based on the data of X-ray diffraction from DNA fibers. This kind of data does not allow unambiguous reconstruction of a molecule structure. One has to design a hypothetical structure, calculate its diffraction pattern, and compare it with the diffraction pattern. If the calculated pattern does not match the experimental data, a different structure has to be designed. In this way, one can, eventually, find a good solution. Using the experimental data of Franklin, Watson and Crick were the first who suggested the correct DNA structure. It was already known at that moment that DNA is the molecule that carries and transmits hereditary information. Therefore, everybody knew that

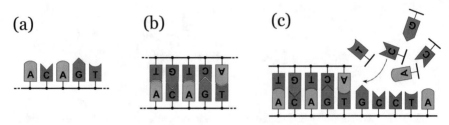

Fig. 1.6 Recording and replicating genetic information. (**a**) The information is coded in the linear string of four different elements (*bases*), A, C, G, and T, which are the side groups of the single-stranded DNA chain (shown by the black line). (**b**) There is a strict correspondence between the sequences of the strands in the double-stranded DNA since A in one strand can be paired only with T in the other strand and G can be paired only with C. Thus, the sequence of the elements in one strand completely specifies the sequence in the other, complementary strand. (**c**) During the *replication* process (making two copies of DNA molecules from one), DNA strands are separated, and the complementary strands are built on each of the parental strands. For simplicity, the synthesis of only one daughter double-stranded DNA is shown

finding the DNA structure is crucial for biology. Still, the structure itself exceeded expectations because it immediately explained the physical principle of inheritance. We will consider this structure in the next subsection, after explaining the principle of recording and duplicating the hereditary information on the simple diagrams.

The DNA structure represents a double helix formed by two single-stranded polynucleotide chains (the repeating units of the chain are called nucleotides). The strands are kept together by weak interactions and can be separated by various means, such as interaction with other molecules or elevated temperatures. The backbone of each chain consists of identical repeating units. Each repeating unit has a side element, A, C, G, or T (Fig. 1.6a). The hereditary, or genetic, information is coded by the sequence of these elements, similar to the strings of letters we use to record words and sentences.

Thus, in principle, the genetic information is written in a very natural way, as a sequence of four letters. We consider later in this chapter conversion of this information into the structure of proteins, which is a key biological process. There is another critical task for the cell, however. The DNA molecule (or molecules) has to be duplicated before the cell division, so one copy of each molecule goes to each of the dividing cells. Remarkably, the physical principle of the latter process immediately follows from the structure of the double helix.

The chains of the double-stranded DNA are kept together by many noncovalent bonds. Simultaneous formation of these bonds is possible because the surface of one chain complements a similar surface of the other chain (Fig. 1.6b). This complementary exists only if A in one strand is paired across the helix with T in the other strand, and G is paired with C. Thus, only AT and GC base pairs can be incorporated into the structure without its distortion, and only these pairs exist in the double-stranded DNA. The complementarity rule means that the sequence of elements in one strand completely specifies the sequence in the other strand.

This principle of DNA structure offers a simple and very elegant principle of duplicating genetic information. First, the parental strands of the double helix have to be separated. Then, the new complementary strands have to be synthesized from

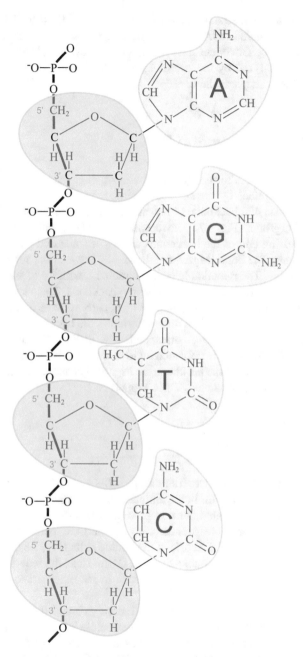

Fig. 1.7 Chemical structure of ssDNA. Each repeating unit of the molecule backbone (shown by bold black lines) consists of 2′-deoxyribose (shaded by pink) and a negatively charged phosphate residue. The side chains (shaded by light blue) are represented by one of four bases: adenine (A), guanine (G), thymine (T), or cytosine (C). Adenine and guanine belong to the group of purines, while the smaller thymine and cytosine are pyrimidines. Grey digits, 3′ and 5′, show the standard numbering of the backbone carbon atoms

the nucleotides on each strand of the parental double helix (Fig. 1.6c). The new daughter double-stranded DNA molecules obtained in this way are identical to the parental double helix. Of course, the synthesis of the new strands does not go spontaneously, it is catalyzed by the coordinated work of many special *enzymes* (see Chap. 6), but the principle is there.

1.3.2   Chemical Structure of DNA

The DNA double helix is formed by two chains of single-stranded DNA (ssDNA), which are held together by noncovalent interactions between their components. Each chain consists of a backbone of repeating units and bases that are attached to each unit as side groups (Fig. 1.7). The repeating unit of the chain is called a *nucleotide*. Each repeating unit of the backbone consists of sugar and phosphate and has six skeletal bonds. The bond lengths and the angles between the adjacent bonds do not change notably. The remarkable conformational flexibility of ssDNA is due to six rotation angles in each repeating unit of the backbone. The backbone has clear directionality, and the method of the numbering of carbon atoms of the sugar, shown in Fig. 1.7, identifies $3'-5'$ or $5'-3'$ directions along the backbone (from bottom to top and from the top to bottom, correspondingly, in Fig. 1.7). It is common to assume a $5'-3'$ direction of the polynucleotide chain when presenting the sequence of bases. The phosphates of the backbone are negatively charged at usual conditions (see Fig. 1.7), and this affects many DNA properties.

1.3.3   The Double Helix

DNA double helix is formed by two strands of ssDNA bound together by noncovalent interaction. The strands are in the antiparallel orientation. The most important structural feature of the double helix is base pairing, the formation of hydrogen bonds by bases from opposite strands. Only AT and GC pairs can be smoothly incorporated into the regular structure of the double helix (Fig. 1.8), and only these pairs appear, normally, in the double-stranded DNA. Thus, the sequence of bases in one strand completely determines the sequence in the other strand. This is a key property for DNA biological functioning.

There are two hydrogen bonds in the AT base pair and three in the GC pair (Fig. 1.8). The base pairs do not just have very close dimensions, but also their external geometries related to the backbones are nearly identical. Therefore, in either of the two orientations (AT or TA and GC or CG), the base pairs can be very well incorporated in a uniform helical structure of DNA.

DNA double helix is right-handed. The base pairs are located inside the helix, while the backbones are at the helix exterior (Fig. 1.9). The planes of the base pairs are nearly perpendicular to the helix axis. As the backbones are winding around the helix axis, the base pairs are rotating around the axis as well. There are 10.5 base

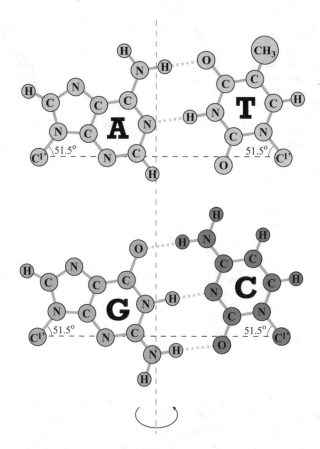

Fig. 1.8 The complementary base pairs. The hydrogen bonds are shown by dashed lines. The bonds connecting bases to $C^{1'}$ atoms of the deoxyribose are also shown. The distances between these bonds and the angles between them are the same for both base pairs. Either of the two orientations of the base pairs, AT or TA and GC or CG, fits well with the structure of the double helix. This follows from the fact that the rotation of a base pair around the axis located in the plane of the base pair (shown by the vertical dashed line) by 180° does not change the position of N–$C^{1'}$ bonds. Therefore, the backbones maintain identical conformation regardless of the DNA sequence. For convenience, each base is shown in a different color (not in all formats)

pairs (bp) per one turn of the helix. The external diameter of the helix is approximately 2.0 nm. The adjacent planes of the base pairs are separated by 0.34 nm along the helix axis. The structure of the double helix allows a massive weak interaction between the bases of adjacent base pairs, called the stacking interaction. This is the most important interaction stabilizing the helix structure.

It is important to emphasize once again that the strands of the double helix are kept together by weak noncovalent interactions. As a result, the complementary strands of the double helix can be easily separated by proteins that interact with DNA, to make the bases accessible for various biochemical processes. We will see many examples of such processes ahead. In general, DNA can exist as a single-stranded polymer, although this form is rarer.

Fig. 1.9 A diagram of DNA double helix. The backbones of both strands (shown by grey) are at the exterior of the helix, while the base pairs (shown by the colors used in Fig. 1.8) are inside, closer to the helix axis. The planes of the base pairs are nearly perpendicular to the helix axis. The helix has two grooves, the minor and the major. The major groove is wider than the minor one. There are no colors in this figure in some formats

1.4   RNA Molecules

Chemically, RNA (ribonucleic acid) is very similar to DNA. The deoxyribose of DNA is replaced by ribose in RNA, where one of two hydrogens attached to the 2′ carbon atom is replaced by the OH group. The second difference is that thymine is replaced by a similar base, uracil (U). A fragment of single-stranded RNA containing all four bases is shown in Fig. 1.10. Although the chemical differences between RNA and DNA seem very small, they are sufficient for many proteins interacting with the nucleic acids to distinguish one from another. RNA molecules can form the double helix similar to DNA, with AU and GC base pairs, but long double-stranded RNA molecules are relatively rare in life. The majority of RNA molecules are single-stranded. Depending on their sequence, some of them can form hairpin-like short double helices (Fig. 1.11). In certain cases, the interaction between the bases can shape the well-defined 3D structure of RNA molecules, as we will see below.

RNA molecules play various roles in cell life, first of all, in transferring the genetic information recorded in DNA to the structure of proteins.

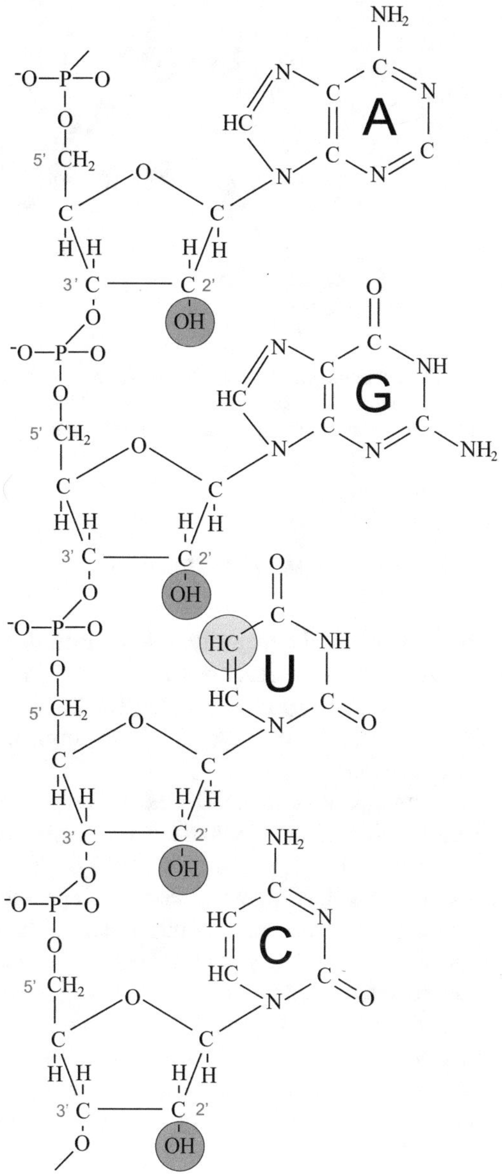

Fig. 1.10 Single-stranded RNA. Places, where RNA molecules differ from DNA, are marked by shaded circles. First, in RNA molecules, the deoxyribose of DNA is replaced by ribose that has a hydroxyl group instead of one of the hydrogens at the 2′ position. Also, thymine is replaced by uracil, a similar base that only lacks the methyl group

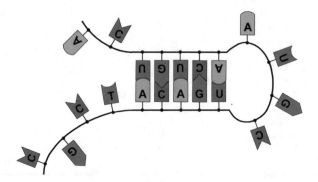

Fig. 1.11 Diagram of a hairpin structure formed by an RNA chain. The structure has a segment of the double helix consisting of five base pairs, a hairpin loop, and segments of unpaired nucleotides at the ends. Large single-stranded RNA can form many such hairpins

1.5 Proteins

1.5.1 The General Principles

Proteins perform the majority of key functions in the cell. Nearly all enzymes that catalyze biochemical reactions are proteins. Proteins serve as regulators of the great majority of cellular processes. They work as molecule motors. Proteins are building elements for many cell structures. They form channels and pumps incorporated into the cell membrane.

Proteins, similar to nucleic acids, represent linear chains of structural units. There are 20 different kinds of units in proteins. As we will see below, the sequence of these units in each protein is coded by the corresponding segment of DNA, a *gene*. The cell has a special complex apparatus that synthesizes protein chains according to the sequences of nucleotides in their genes. A key property of the proteins is that each newly synthesized chain spontaneously folds into a unique 3D structure determined by the sequence of structural units in this chain. The structure is stabilized by weak interactions between these units. It is this ability of protein chains to fold into specific 3D structures that makes life possible.

It is also critically important that all varieties of proteins is produced by the same cellular apparatus, which only synthesizes linear chains by adding one structural unit after another to the growing polypeptide. The choice of the particular unit is specified by the nucleotide sequence in the protein gene. Thus, the linear chemical structure of the polypeptides enormously simplifies their synthesis. It also greatly simplifies the coding of the protein 3D structure—only information on the sequence of units in the protein chain is needed.

The fact that the same apparatus is used to synthesize all proteins is a critical property of living cells. In principle, various functions could be performed by other macromolecules of a different, nonprotein, nature. We know, for example, that many complex antibiotics produced by microbial cells have a nonprotein nature.

The synthesis of such antibiotics requires many unique enzymes, which are macromolecules themselves. The cells can use such sets of unique enzymes only for the production of a very limited number of needed macromolecules, but it is unacceptable as a general solution. Indeed, if the production of each enzyme requires a few other unique enzymes, the total number of needed enzymes would go to infinity. The only visible solution here is to use the same set of enzymes to produce all, or nearly all, macromolecules. We know now that this is possible due to the remarkable property of polypeptide chains.

Let us now consider the properties of proteins in more detail.

1.5.2 Structure of Proteins

From the chemical point of view, proteins are linear polymer chains consisting of different units, *amino acids*. Protein chains, called *polypeptides*, contain from a few dozen to 2000 amino acids. The backbone of proteins consists of identical repeating units that have certain directionality (Fig. 1.12). Each amino acid adds three single bonds in the backbone. Rotation is possible around two of these bonds, C–C and N–C, and this provides very high conformational flexibility of the chain. Each amino acid has its unique side group, which specifies its properties. Twenty different amino acids are used by cells to build proteins, and the properties of these amino acids vary greatly. Some of them are negatively charged; others carry positive charges. They have different sizes. Some of them are polar (whose side chains are capable of forming hydrogen bonds with other atoms), while others are nonpolar. The atomic structures of a few amino acids and their main properties are given in Table 1.1. As we mentioned above, the 3D structure of a protein is stabilized by noncovalent interactions between the amino acids of the polypeptide. These interactions are different in each structure. In general, proteins fold into structures that have minimal free energy.[2] The structures of proteins determine their biochemical properties and functioning in the cell.

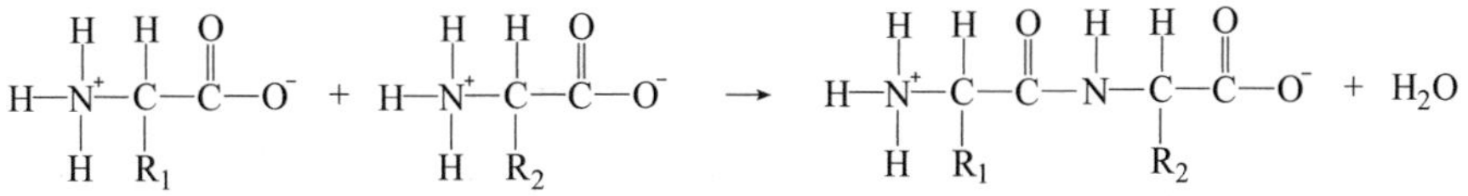

Fig. 1.12 Formation of a dipeptide by two amino acids. R_1 and R_2 are side groups. Under neutral pH, the N-terminal end of polypeptides is positively charged, while the C-terminal end is negatively charged

---

[2] The free energy, G, is defined as $G = E - TS$, where E is the energy of the considered system, S is the entropy of the participating compounds, and T is the absolute temperature. The value of S changes in chemical reactions and conformational transformations. The formation of a unique 3D structure by a protein chain reduces its entropy, but in a stable structure this reduction is compensated by a decrease in E due to numerous weak interactions between the atoms.

Table 1.1 Some amino acids which appear in proteins and their RNA codons

Side group of amino acid	Three-letter notation	Type	Chemical structure of the side group	Codon
Alanine	Ala	Nonpolar	$-CH_3$	gca, gcc, gcg, gcu
Glycine	Gly	Nonpolar	$-H$	gga, ggc, ggg, ggu
Histidine	His	Positively charged	$-CH_2-\overset{CH-NH^+}{\underset{NH-CH}{\langle\parallel}}$	cac, cau
Aspartic acid	Asp	Negatively charged	$-CH_2-C\overset{\nearrow O}{\searrow O^-}$	gac, gau
Asparagine	Asn	Uncharged polar	$-CH_2-C\overset{\nearrow O}{\searrow NH_2}$	aac, aau
Tyrosine	Tyr	Uncharged polar	$-CH_2-\langle\bigcirc\rangle-OH$	uac, uau

The enormous amount of polypeptide sequences can be made from 20 different amino acids. For example, for a polypeptide of 100 amino acids, there are 20^{100} different possibilities. However, only a very small fraction of the total amount of polypeptides folds into stable and unique 3D structures. A great majority of polypeptide chains of proteins have this property, which is necessary for their function. The sequences of proteins were selected by the evolution from the huge amount of possibilities, and we know that proteins perform their tasks with amazing efficiency and precision. It does not mean, however, that a better protein for a particular task cannot be designed.

The diversity of amino acids allows the making of protein molecules with a nearly infinite variation of properties. There are proteins that recognize and bind other molecules with very high specificity and affinity. Proteins work as enzymes and can catalyze a huge variety of biochemical reactions. In some cases, these reactions are very complex multistep processes and proteins work like molecular machines, as we will discuss in subsequent chapters. Although many various tasks can be solved by nuclear acids, the capabilities of polypeptide-based molecules are enormously larger. The last column of the table shows nucleotide codes of the corresponding amino acids (see Sect. 1.6).

It is now easy to determine the sequence of amino acids in a particular protein. It was much more difficult, until very recently, to obtain its 3D structure, the task took months and sometimes years of work. Still, a 3D structure is needed to understand how a protein accomplishes its function. Over a few decades, in hundreds of laboratories around the world, researchers worked hard to determine these structures. About 180,000 3D structures of proteins have been determined by X-ray analysis and by nuclear magnetic resonance (NMR). At the same time, many researchers tried to obtain the structures by computer simulation of the protein folding, using the sequence of amino acids in the polypeptide chain. The computational approaches were based on the calculation of the energy of all interactions between the atoms for each conformation of the chain obtained in the simulation. However, the accuracy of energy calculation was insufficient, and the approach has brought only very limited success.

The unexpected breakthrough came from another direction. In 2016 a group of researchers decided to apply an approach based on artificial intelligence to the problem of protein folding. Artificial intelligence is a computer program that is improving its predicting power by estimating the previous results and correcting itself. Of course, the application of the approach to the problem of protein folding became possible only because about 200,000 3D structures of various proteins were available from the experimental studies. A comparison of the program's earlier predictions with the experimentally determined structures was used in its "self-learning." Over the next few years, it became clear that the approach is capable of solving the problem with remarkable accuracy and efficiency. In the second half of 2022, the leading group of researchers determined the 3D structures of nearly all proteins whose sequence had been known. More than 200 million predicted structures have been deposited into the open-access database so that everyone can now use them in research.

Analysis of obtained 3D structures of proteins, which contain thousands of atoms, is also hardly possible without special computer programs. Structural analysis of proteins is simplified by the fact that two common structural motifs are present in the great majority of proteins. The first motif is *α-helix*. It represents a right-handed helix with 3.6 amino acids per helix turn. It is stabilized by hydrogen bonds between N–H and O=C groups of the polypeptide backbone (Fig. 1.13a). The amino acid side groups are located outside the α-helix and do not affect the helix structure (Fig. 1.13b). The second motif is the β-sheet. It is also stabilized by hydrogen bonds between the same groups of the backbone (Fig. 1.13c). The conformation of the backbone in this motif forms a nearly flat surface. The side groups of the amino acids are located below or above the surface (Fig. 1.13d), so the structure of β-sheets is independent, in a good approximation, of the side groups of included amino acids. Although only atoms of the polypeptide backbone participate in the hydrogen bonds that stabilize α-helices and β-sheets, the side groups affect the propensity of various amino acids to be a part of either of the structural motifs.

It is common, in the drawing of proteins, to replace α-helices and β-sheets with ribbons (Fig. 1.14). Although such representation skips many details, some of which are very essential, it gives a good general image of the protein structure. Of course, 3D structures of proteins have many other not-so-regular segments. The total fraction of such irregular segments, as well as the fractions of the segments in the α-helices and β-sheets, varies greatly among different proteins. In general, each amino acid may contribute to the structure's stability by interacting with other amino acids, water molecules, ions, and other small molecules. Four structures of proteins shown in Fig. 1.15 illustrate the variety of protein structures.

One more property of proteins is fundamental for their biological role. The structures of proteins can be easily changed in the course of their functioning. Such changes are possible only because the structures are stabilized by weak interactions between chemical groups of proteins, so the changes do not require much energy. Global structural changes can be caused by binding another molecule to a specific

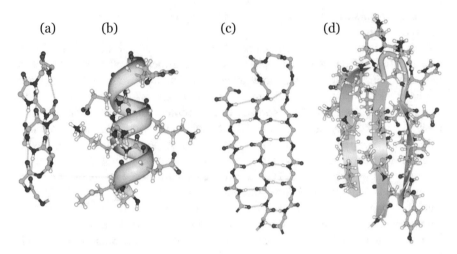

Fig. 1.13 The major structural motifs of proteins, α-helices and β-sheets. Only atoms of the poly-peptide backbone participate in the networks of hydrogen bonds (shown by dashed lines) that sta-bilize these structural motifs (**a, c**). For clarity, the amino acid side groups and hydrogen atoms that are not participating in the hydrogen bonding are not shown in panels (**a**) and (**c**). The amino acid side groups are located outside these patterns of the backbone (**b, d**). Panels (**b**) and (**d**) also con-tain the ribbon representation of the backbone for α-helices and β-sheets, which are widely used in drawing the protein structures (see below). In the color formats of the figure, the carbon, oxygen, nitrogen, and hydrogen atoms are shown in green, red, blue, and gray, correspondingly. The images from RCSB PDB were obtained with the Molecular Biology Toolkit (Moreland et al., 2005, *BMC Bioinformatics*, 6:21)

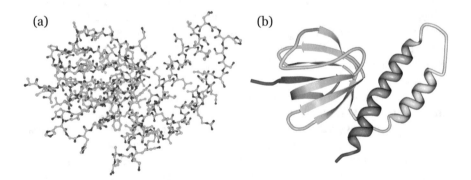

Fig. 1.14 Two ways of presentation of the protein structures. (**a**) The image shows the structure of the epsilon subunit of the F1-ATPase (Wilkens & Capaldi, *J. Biol. Chem.*, 1998. **273**, 26645–51; PDB ID: 1BSN) in all-atom presentation (without hydrogen atoms). (**b**) The same structure is shown in the ribbon presentation. The thin non-ribbon segments of the chain do not belong to either α-helices or β-sheets. The images from RCSB PDB were obtained with the Molecular Biology Toolkit (Moreland et al., 2005, *BMC Bioinformatics*, 6:21)

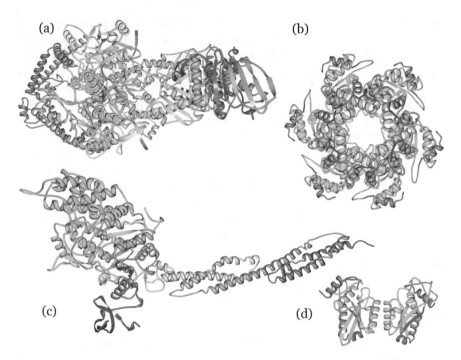

Fig. 1.15 Diversity of protein structures. (**a**) DNA topoisomerase (Schmidt et al., *Nat. Struct. Mol. Biol.*, 2012. **19**, 1147–54; PDB ID: 4GFH) catalyzes the passage of one segment of double-stranded DNA through another. (**b**) Helicase (Enemark, E.J. and L. Joshua-Tor, *Nature*, 2006. **442**, 270–5; PDB ID: 2GXA) is the protein that moves along the double helix and unwinds it. The protein consists of six identical subunits. (**c**) Motor protein non-muscle myosin (Chinthalapudi et al., *PDB*, 2016; PDB ID: 2YCU). (**d**) Restriction endonuclease Bam HI (Aggarwal & Newman, *Structure*, 1994. **2**, 439–52, PDB ID: 1BAM), the enzyme that binds DNA segments with the sequence GGATCC and cuts both strands. The protein consists of two identical subunits. The images from RCSB PDB were obtained with the Molecular Biology Toolkit (Moreland et al., 2005, *BMC Bioinformatics*, 6:21)

site of the protein. This molecule can be a substrate for a chemical reaction catalyzed by the enzyme, *adenosine triphosphate* (ATP), molecule that serves as an energy source for various biochemical processes, or another macromolecule (see Fig. 2.8).

1.6 Genetic Code

From the chemical point of view, proteins are linear chains of amino acids. Therefore, all information that is needed to reproduce a particular protein is in the sequence of amino acids in the polypeptide chain. The folding of the polypeptide into a unique 3D structure and the potential function of the protein are completely specified by this sequence. The sequence of amino acids can be written as a string of letters, and this is exactly how it is written in DNA molecules. Of course, there are 20 different amino acids in proteins and only four different bases in DNA. So, more than one nucleotide is required to code a particular amino acid. There are $4 \times 4 = 16$ different two-letter words that can be made of four different letters and $4 \times 4 \times 4 = 64$ different three-letter words. So, to code 20 amino acids, at least a three-letter code is needed. Therefore, it is not surprising that nature used a triplet code to record protein sequences by DNA bases. More than one different triplet (codon) codes each amino acid (except methionine and tryptophan, which are coded by single codons). The genetic codes for some amino acids are presented in Table 1.1. The codons shown in the table are RNA codons since RNA molecules are used as intermediates to transfer genetic information from DNA to proteins (discussed below in this chapter). There are also three STOP codons, uaa, uag, and uga. These codons terminate the synthesis of a protein chain. There is one special codon, aug, which codes methionine. The codon is used for the start of the protein chain. An additional mechanism is used to distinguish starting aug codons from the same codons internal for genes, which should not be used as the starting points. A segment of DNA that codes the amino acid sequence of a protein is the *gene* of this protein.

One important feature of the genetic code is that codons are not separated one from another in the genes. Thus, it looks like a regular English text where all spaces between the words were deleted. As a result, any DNA segment can code, in principle, three different polypeptide chains which are obtained by shifting the *reading frame* (Fig. 1.16). Of course, each coding segment has to start from the aug codon, and this sets the correct reading frame.

Interestingly, some viruses (see Sect. 11.1)) use the frameshift to code different proteins by overlapping stretches of DNA. This overlapping allows them to have a smaller total DNA length, which is important for their proliferation.

As we wrote above, the genetic code is universal for all cells on our planet.

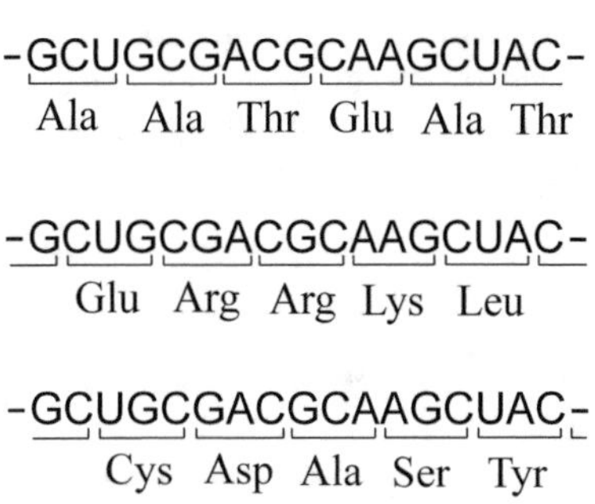

Fig. 1.16 Shifting the reading frame in the nucleotide sequence completely changes the sequence of amino acids coded by the same RNA segment

-GCUGCGACGCAAGCUAC-
Ala Ala Thr Glu Ala Thr

-GCUGCGACGCAAGCUAC-
Glu Arg Arg Lys Leu

-GCUGCGACGCAAGCUAC-
Cys Asp Ala Ser Tyr

1.7   From DNA to Proteins

The information on protein sequences is recorded in DNA, and knowledge of the genetic code allows us to easily convert a sequence of DNA bases into a sequence of amino acids. It is not a simple task for the cell, however, to perform the synthesis of proteins following the nucleotide sequences of genes. There is no direct chemical fitting between the three-letter codons and amino acids, similar to the complementarity between the bases of the double helix. Therefore, there is no straightforward way to read information coded in DNA molecules and use it for protein synthesis. Protein synthesis is a complex process with an elaborate system of regulation, and a large part of this book will be dedicated to it. Here we will only very briefly outline its major steps and elements.

Proteins are not synthesized on DNA molecules directly. Instead, there is an intermediate step in the process, called *transcription*. During the transcription, the cell synthesizes RNA molecules that correspond to individual genes (or groups of genes) on the corresponding segments of DNA. These RNA molecules are called *messenger RNA* (mRNA). The synthesis starts at special points of DNA and is performed by a set of proteins, among which the major one is *RNA polymerase*. The starting points are located upstream of the genes. The RNA polymerase unwinds a few base pairs of DNA, exposing the corresponding bases. One strand of DNA is used as a template for the synthesis of mRNA. The RNA polymerase catalyzes the extension of the mRNA chain by one nucleotide at each step of the synthesis. The ribonucleotide incorporated into the growing mRNA chain is specified by the same rules of complementarity used in the DNA structure, with the exception that thymine is replaced by uracil in the RNA chain. When a ribonucleotide is added to the growing RNA chain, the enzyme moves to the next base pair of DNA. Simultaneously, it opens one more DNA base pair in front of itself (Fig. 1.17). The ribonucleotides needed for the synthesis are taken from the surrounding solution. The synthesis creates hybrid DNA/RNA double helix. Behind the moving RNA polymerase, the end of the growing mRNA dissociates from the DNA strand, and DNA strands reassociate. When the RNA polymerase reaches the end of the transcription segment, the single-stranded mRNA is released.

It is convenient for the cell to have mRNA molecules as intermediates in protein synthesis. Depending on the changing need in a particular protein, the corresponding mRNA can have a larger or smaller number of copies in the cell. This number can be increased or reduced with time because mRNA molecules are not only permanently synthesized inside the cell but are also digested. Although molecules of mRNA can consist of a few thousand nucleotides, they are, on average, many times shorter than DNA molecules that code them.

Thus, in transcription, the first stage of protein synthesis, the needed information is transferred from the DNA to the nucleotide sequence of mRNA. Transcription, however, does not simplify converting the genetic information to the sequence of amino acids of proteins. There is no complementarity between codons of mRNA and the amino acids, and the cell has to use special adaptors in the process of protein

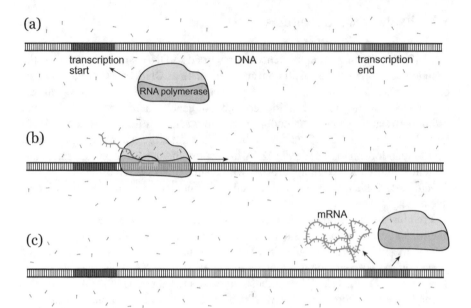

Fig. 1.17 Transcription of a gene by RNA polymerase. The process produces mRNA that codes the gene (or RNA molecule used by the cell for other needs, as it is described below). (**a**) The process starts when the RNA polymerase binds to the DNA site with the special sequence, the starting point of the transcription. (**b**) The enzyme unwinds a few DNA base pairs and starts RNA synthesis by using the coding DNA strand as a template. After the needed ribonucleotide (shown as small particles scattered around) from the surrounding solution binds to the corresponding DNA nucleotide, the enzyme attaches it to the growing mRNA chain. At the end of each step, the RNA polymerase moves to the next base on the template. (**c**) The RNA polymerase and newly synthesized mRNA dissociate from DNA when the enzyme reaches the gene end, a DNA site with a special sequence

synthesis, called *translation*. The adaptors represent a class of special RNA molecules, which are called *transport RNAs* (tRNAs). Each tRNA molecule can carry its specific amino acid and bind to the corresponding anticodon. tRNA molecules have about 80 nucleotides and fold into very compact specific structures. A substantial length is necessary for these molecules to set up their specificity both to the codons of mRNA and to the corresponding amino acid. Thus, the number of different tRNA molecules corresponds to the number of various codons. Codon specificity to tRNA molecule is set by the segment of three nucleotides (anticodon), which sequence has complementarity to the sequence of the corresponding codon. Each tRNA can carry the corresponding amino acid. The specificity of a tRNA molecule to particular amino acid is set by a few elements of a tRNA structure, and the sequence of its anticodon is one of them. Still, the molecules of tRNA do not recognize the needed amino acids themselves. The attachment of an amino acid to its specific tRNA is performed by special enzymes, aminoacyl-tRNA synthetases, that catalyze the formation of the covalent bond between the amino acid and corresponding tRNA. For each amino acid, the cell has a specific aminoacyl-tRNA synthetase that recognizes this amino acid and the corresponding tRNA (Fig. 1.18).

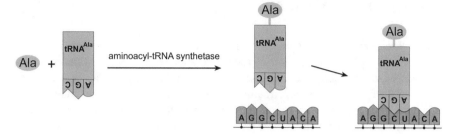

Fig. 1.18 Formation of the aminoacyl-tRNA and its binding to mRNA. In this process, the amino acid, alanine in the shown case, is covalently attached to tRNA^{Ala} by the alanine-specific aminoacyl-tRNA synthetase. The enzyme can only catalyze the bond formation between alanine and tRNA^{Ala}. The tRNA^{Ala} with attached alanine, obtained in this way, is capable of binding to the corresponding codon, GCU, of mRNA

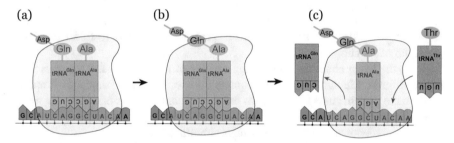

Fig. 1.19 Protein synthesis on the ribosome. (**a**) Two aminoacyl-tRNAs inside the ribosome (shown by yellow) interact with the corresponding codons of mRNA (light brown), forming complementary base pairs. The aminoacyl-tRNA^{Ala} (tRNA carrying the amino acid alanine) is not bound yet to the growing polypeptide chain. (**b**) The covalent bond between tRNA^{Gln} and amino acid glycine is replaced by the covalent bond between glycine and alanine. The aminoacyl-tRNA^{Ala}, mRNA, and the discharged tRNA^{Gln} are shifted left in the ribosome. (**c**) tRNA^{Gln} is released from the complex. New aminoacyl-tRNA can now bind the vacated position in the ribosome. The ribosome is not shown in scale in this drawing; its actual size is many times larger

The amino acids with the attached tRNA molecules form substrates for protein synthesis. This synthesis is processed in the *ribosome*, a very large and complex catalytic machine consisting of more than 50 proteins and a few *ribosomal RNAs*, rRNAs. mRNA bound with a ribosome is used as a template in the synthesis. The general diagram of the process is shown in Fig. 1.19. The ribosome binds a molecule of mRNA and catches, one by one, the needed molecules of tRNA with the attached amino acids (*aminoacyl-tRNA*) from the surrounding solution. At each step of the synthesis, the choice of aminoacyl-tRNA is determined by the mRNA codons. When the needed aminoacyl-tRNA is properly placed in the ribosome, the covalent bond between tRNA and the amino acid is replaced by the covalent bond between this amino acid and the growing end of the polypeptide. The discharged tRNA is then released from the ribosome. The synthesis ends when the stop codon is reached, and a new protein is released from the ribosome.

1.8 Concluding Remarks

We outlined here the major processes of life, storage of the genetic information in DNA and its replication, transferring the information to mRNA called transcription, and mRNA-dependent protein synthesis called translation. The way of converting genetic information to protein structures involves many enzymes and adaptors and it is a complex process with very sophisticated regulation. This key process of life, allowing the cell to produce large protein molecules of enormous structural and functional diversity, is amazing in its efficiency. Let us summarize once again its major features.

A huge variety of protein structures are needed for various functions of the cells. Proteins serve as enzymes catalyzing thousands of different biochemical reactions. They serve as molecular motors and building blocks. They form *ion channels* and *membrane pumps* incorporated in the cell membrane. They transmit signals in thousands of signaling pathways. Large special protein molecules of the immune system, antibodies, bind nearly any harmful molecule with great specificity and affinity, to protect the cells. Two key properties of proteins allow them to fulfill this huge spectrum of functions.

First, the linear chains of proteins are capable of folding into an enormous variety of different structures, which are completely specified by the sequence of amino acids in the protein chains. This means that a very complex task of coding 3D structures of proteins is reduced to a simple task of coding linear sequences of amino acids in polypeptides. This is an enormous simplification!

Second, since the structures of proteins are stabilized by weak noncovalent chemical bonds, these structures can be changed during protein functioning. Without this structural flexibility, the proteins simply could not perform the majority of their functions. Structural flexibility is a common feature of nearly all biological macromolecules, including nucleic acids. This property is similar to the property of human-made machines that all have moving parts.

These two properties are necessary for life. It may be possible that life on some other planet is based on polymer chains consisting of another set of monomers rather than amino acids. However, whatever would be the monomers, the corresponding polymers must have both of these two properties. The same is probably true for a very important feature of protein synthesis inside the cells.

Protein synthesis in living cells is incredibly efficient from a technological point of view. Because all proteins are synthesized as linear chains of amino acids, the cells have to have only one kind of manufacturing equipment for the synthesis of all proteins. Such universality enormously simplifies production and allows cells to switch from the synthesis of one protein to another without any changes in the manufacturing equipment. This key feature of protein synthesis greatly simplifies the general design of living cells.

Chapter 2
Enzymes, Conformational Changes, Energy, and Molecular Motors

2.1 The Motion of Macromolecules in Water

Molecules in a water solution are in constant random motion, and the cell is not an exception. The smaller molecules move faster, while the larger ones move slower. This random motion involves the translational displacement of molecules, their rotation, and changing of their shape if the molecules are flexible. A key feature of this random motion is that at every moment of time, a molecule does not remember about the direction and speed of its movement in the previous moment. This is very different from the movement of bodies in our macroscopic world, where bodies have substantial inertia, which is the tendency to maintain the direction and speed of their movement. If a body is smaller, its inertia is smaller, and less effort is required to change the direction and speed of the movement. If we go down to bodies of molecule size, their inertia, specified by the body's mass, is negligible in water solutions. Why is it so? The laws of physics that govern the motion of molecules and macroscopic bodies are the same. The difference is due to the viscosity of water which is huge in the scale of molecules. The viscosity of water creates a friction force that is opposite to the direction of the movement, so it tends to stop a moving body. This force is proportional to the linear size of the body. The body mass, however, is proportional to its volume, or the third degree of the linear size. So, when the body's size is diminishing, the inertia is decreasing much faster than the friction force. Suppose that we have a ball of 10 mm in diameter and a microscopic spherical particle of the same density with a diameter of 10 nm (the size of a large protein), so their diameters differ by a factor of 10^6. Suppose the two move in water at the same speed. The friction force will be 10^6 times smaller for the second sphere, while its inertia is 10^{18} times smaller! If there is no thermal motion of water molecules, a particle of molecular size would be stopped by the friction force nearly immediately. Instead, it is permanently pushed by water molecules and permanently changes the direction of its movement. This kind of motion is called diffusion. In the scale of macromolecules, diffusion can be superimposed with

A. Vologodskii, *The Basics of Molecular Biology*,
https://doi.org/10.1007/978-3-031-19404-7_2

directed motion in the force field, such as the movement of charged molecules in the electric field. However, inside the cell, diffusion is the major type of motion. Remarkably, the cells know how to convert diffusion into directed motion. We will consider this later in the chapter.

Structural changes in the macromolecules also occur as a diffusion, intramolecular diffusion in this case. Usually, only structural changes of the molecule backbone are considered in this context, and they are called *conformational changes*. We will use the term below. Usually, the conformational changes result from chaotic thermal motion of the macromolecule parts. Still, many conformational changes occur sufficiently fast, even if they are associated with a certain increase in the energy of the macromolecule. Although the average energy of molecules is specified by the solution temperature, random fluctuations of this energy are large. So, the actual energy of the molecule continually changes, and at each moment, it can be larger or smaller than the average value. This makes possible conformational changes in the macromolecule which are associated with an increase in its energy. The ability of the macromolecules to diffuse into conformations with higher energy is critically important for the functioning of many biological macromolecules.

2.2   Enzymes

Life is not possible without enzymes and biological catalysts, which increase the speed of biochemical reactions by many orders of magnitude. Although like any catalysts, they cannot change the chemical equilibrium, they can catalyze energetically unfavorable reactions. Enzymes achieve this by coupling such reactions with the reactions that supply energy. They can catalyze multistep processes that could not occur spontaneously at a detectable rate, working like very complex molecular machines. Due to their crucial role in life, it is worth considering enzymatic catalysis in some detail.

The majority of enzymes are globular proteins (Fig. 2.1). Their active center, where they bind the substrate (the compound which participates in the reaction), can occupy a small fraction of the enzyme volume.

According to the major law of thermodynamics, compounds that have higher energy should more often undergo transitions into compounds with lower energy rather than the reverse transitions. In many cases, however, even the transitions associated with the reduction of energy do not occur at a notable rate. This is because any chemical reaction has to go through the *transition state*, which is associated with the energy barrier, called the *activation energy* E_a. Let us consider a simple reaction where reaction product AB is formed by joining reactants A and B:

$$A + B \rightarrow AB.$$

The change of energy in this reaction is illustrated in Fig. 2.2. The barrier in the energy profile slows down the reaction rate dramatically. To overcome this barrier, the molecule has to receive energy that exceeds the value E_a. For the considering example, this means that reactants A and B have to collide at high speed. This speed

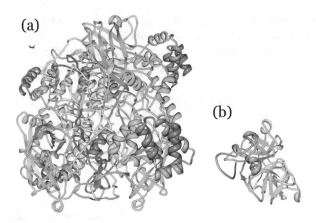

Fig. 2.1 Structures of two enzymes. (**a**) Catalase (Putnam et al., *J. Mol. Biol.*, 2000. **296**, 295–309; PDB ID: 1DGF) is a complex of four subunits, and each subunit acts independently. The enzyme converts two molecules of hydrogen peroxide, H_2O_2, into two molecules of water, H_2O, and oxygen, O_2. The substrate, shown in the upper left corner, is much smaller than the enzyme. (**b**) Lysine-specific protease (Kitagawa & Katsube, *J. Biol. Chem.*, 1989. **264**, 3832–9; PDB ID: 1ARB). The enzyme cuts the polypeptide chain backbone in positions adjacent to the amino acid lysine. The images from RCSB PDB were obtained with the Molecular Biology Toolkit (Moreland et al., 2005, BMC Bioinformatics, 6:21)

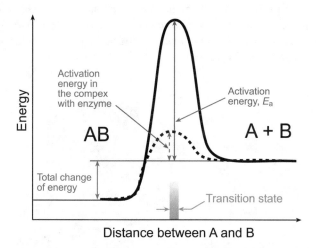

Fig. 2.2 The activation energy and the enzymatic catalysis. The energy of imaginary particles A and B is plotted as a function of the distance between them. The particles can form compound AB whose average energy is lower than the average energy of A and B together. The reaction can occur only if the energy of compounds A and B occasionally exceeds the activation energy E_a, which is usually high. The enzyme that catalyzes the reaction binds molecules A and B in a state which is closer to the transition state. The energy change along the reaction coordinate for the enzyme-bound substrate is shown by the dashed line. The value E_a is greatly reduced in this case, resulting in a dramatic increase in the reaction rate

and, correspondingly, energy can be accumulated in A and B through random collisions with other molecules, which is through the process called thermal motion. However, for chemically stable compounds, this happens very rarely because the average energy of thermal motion at the usual temperature is many times lower than the value E_a.

The reaction rate, k, can be described by the equation

$$k = A \exp\left(-\frac{E_a}{RT}\right),$$

where R is the gas constant, T is the absolute temperature, and A is a reaction-dependent coefficient. If the value E_a exceeds RT by many times, the reaction rate is very low at usual temperatures. For example, for the reaction that converts two molecules of H_2O_2 into O_2 and two molecules of H_2O, the value E_a equals 18 kcal/mol. The value of RT at the usual temperature is close to 0.61 kcal/mol, so

$$\exp\left(-\frac{E_a}{RT}\right) \cong 1.5 \times 10^{-13}.$$

So, it is not surprising that the reaction rate is extremely low at such a temperature. Of course, the rate is growing with the rise of temperature, but physiological conditions are restricted by a relatively small range of temperatures.

Enzymes bind their substrates with exceptional specificity. This specificity is achieved by precise complementarity between the surfaces of the binding site of the enzyme and the substrate. Either a macromolecule or a small molecule can be a substrate for the enzyme (Fig. 2.3).

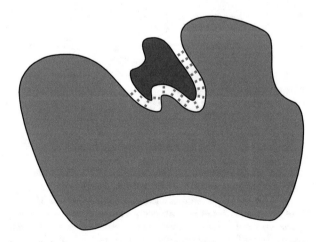

Fig. 2.3 A diagram of a complex between the enzyme and its substrate. The complex is stabilized by many noncovalent bonds (shown as short dashed lines). The formation of all these bonds is possible due to the complementarity of the surfaces of the enzyme (large body) and the substrate (small body)

The energy barrier for the reaction corresponds to the most unstable state of the molecule, called the transition state. We can imagine the transition state as the one where the molecule structure is deformed relative to its normal structure. Enzymes bind their substrates in states which are close to the transition states. As a result, a bound substrate has a lower barrier for the reaction, and this can increase the reaction rate greatly (see Fig. 2.2). In the example with H_2O_2, the enzyme which catalyzes the reaction, *catalase*, diminishes the value of E_a to 5.5 kcal/mol. This corresponds to an increase in the reaction rate by nearly 10^9!

The efficiency of enzymes is amazing. Greatly accelerating the rate of biochemical reactions, they do it with remarkable specificity. They catalyze reactions with their substrates but do nothing with very similar molecules. For example, they can easily distinguish between single-stranded DNA and RNA molecules, although the chemical structures of these polynucleotides differ by a single hydroxyl (see Fig. 1.10).

2.3 Conformational Flexibility of Proteins and Allosteric Transitions

As we discussed in Chap. 1, 3D structures of proteins are specified by the sequences of amino acids in their chains. The structures are stabilized by many weak noncovalent interactions between the protein atoms, and as a result, they are rather flexible. They fluctuate due to thermal motion and can be relatively easily changed by various factors. Some substitutions of a single amino acid may result in a substantial change in a protein structure. The cells use chemical modifications of some amino acids (such as the addition of a phosphate group to an amino acid) to change the properties of proteins. These modifications can inhibit or activate enzymes. The structure of a protein can also be changed by noncovalent interaction with another molecule. Sometimes a molecule interacting with a protein is rather small and the interaction involves only a small fraction of the protein surface. However, the binding of even a small molecule can cause global changes in the protein structure. A change

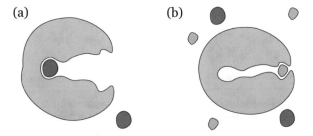

Fig. 2.4 Diagram of an allosteric transition in a protein that serves as an enzyme. In state A, the enzyme (shown in blue) can bind to the substrate (red) in its active site. In state B, the enzyme is bound with another small molecule (green). This binding causes a global change in the protein structure, leaving insufficient space to bind the substrate in the active site

in the global structure of a protein due to the binding of a small molecule or due to a local chemical modification is called *allosteric transition*. The ability of allosteric transitions is a very important property of proteins. The allosteric transitions not only regulate protein activity but also allow proteins to work as molecular machines that catalyze a few successive reactions. A schematic picture of an allosteric transition in a protein caused by the binding of a small molecule is shown in Fig. 2.4.

The key role of allosteric transitions in enzyme action can be well illustrated by the chain of reactions catalyzed by type II topoisomerase. This amazing enzyme catalyzes the passing of one segment of the double helix through another. DNA molecules are so long that topological entanglements between them during cell life are unavoidable (Fig. 2.5). Such entanglements would be a deadly obstacle to cell development and division. The topoisomerases allow the cell to resolve all problems related to the entanglements between DNA molecules. The sequence of the reaction steps and related allosteric changes of the enzyme conformation is described in Fig. 2.6.

The topoisomerases work like a very complex machine capable of performing a chain of reaction steps. Each successive step of the catalytic cycle requires a change of the enzyme conformation and represents an allosteric transition caused by interaction with DNA segments, ATP, and ADP molecules. The first such transition is caused by binding the first DNA segment. In the new state of the enzyme, the gate

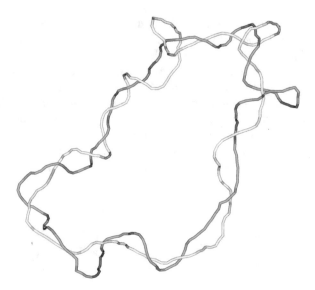

Fig. 2.5 Diagram of linked circular double-stranded DNA molecules. The circular form of DNA is widespread in nature. For clarity, the molecules are shown in different colors. The links of this kind are obtained after the replication of circular DNA. During the cell division, the molecules have to go to different daughter cells, so they have to be unlinked. Special enzymes, type II DNA topoisomerases, solve this problem by catalyzing the passage of one DNA segment through another (Fig. 2.6)

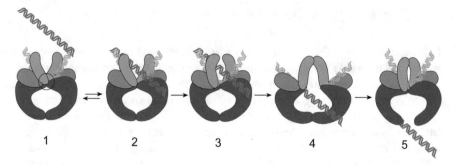

Fig. 2.6 The chain of reactions catalyzed by type II DNA topoisomerases. In state 1, the enzyme is bound with a first DNA segment, which is bent in the complex (shown in yellow). In its open-clamp conformation, the enzyme waits for another DNA segment (shown in light brown). When such a segment gets inside the clamp (state 2), it causes the allosteric transition that closes the clamp (state 3). Closure of the clamp (state 3) triggers breaking the first segment and opening the DNA gate (circled in state 1), allowing the second segment to reach the central cavity of the enzyme (state 4). The latter events trigger the closure of the DNA gate and rejoining the ends of the broken segment. Closure of the DNA gate triggers the last transition in the enzyme, opening the exit gate that allows the release of the second segment from the complex (state 5). The net result of these steps is passing the second DNA segment through the first one

for a second segment is open. Diffusion of the second segment through the gate and its binding with a certain site of the enzyme triggers the next catalytic act, breaking the first segment and attaching the new DNA ends to the enzyme. This creates a gap in the first DNA segment. The second segment can now pass through the gap to the large lower cavity of the enzyme. The allosteric transition following this event results in opening the exit gate for the second segment. Interaction of the second segment with the enzyme in the new state is weakened, so the segment diffuses through the exit gate. The release of the second segment from the enzyme causes the last allosteric transition, resealing the ends of the first segment and releasing it.

It is important to emphasize that the described sequence of the allosteric transitions and reaction steps, catalyzed by the enzyme, is incorporated into its properties. It is possible to extract the enzyme from the cells, purify it, and run the reaction in a tube where only DNA and ATP molecules are added.

There is no change of the energy of DNA molecules in the reaction catalyzed by the topoisomerases, in a good approximation. Therefore, on average, each reaction step can go in either direction. Clearly, under this condition, only a small fraction of the started multistep reactions will be completed, because, at any moment, the sequence of the reaction steps can be reversed. To make the strand-passing reaction more efficient, the enzyme has to keep the sequence of the successive steps shown in Fig. 2.6. It is achieved by coupling one of the reaction steps, energetically unfavorable, with ATP hydrolysis. The coupling makes the desired direction of the entire step energetically favorable and, correspondingly, more probable. We will discuss this issue in detail in the next section.

2.4   The Energy in Cell Life

Life requires energy. The cells need it in many biochemical reactions and processes and for the directed movement of cell components and entire organisms. There are two major ways to obtain the needed energy. Photosynthetic organisms, such as plants, harvest the energy of the sun's electromagnetic radiation. The energy of radiation is used in plants to synthesize sugars, compounds with high-energy chemical bonds. Many organisms, first of all animals, obtain energy by eating other organisms and converting the chemical energy stored in the food into molecules of ATP. Although some other molecules also serve as energy carriers, ATP is the universal currency of energy. We will outline the process of food digestion at the end of this chapter.

The energy stored in ATP is released in ATP hydrolysis, the reaction which gives ADP and inorganic phosphate (Fig. 2.7). ATP hydrolysis is coupled with nearly all processes where energy is needed.

In many biochemical reactions, the free energy of substrates is lower than the free energy of products, so the reactions increase the free energy of the compounds.

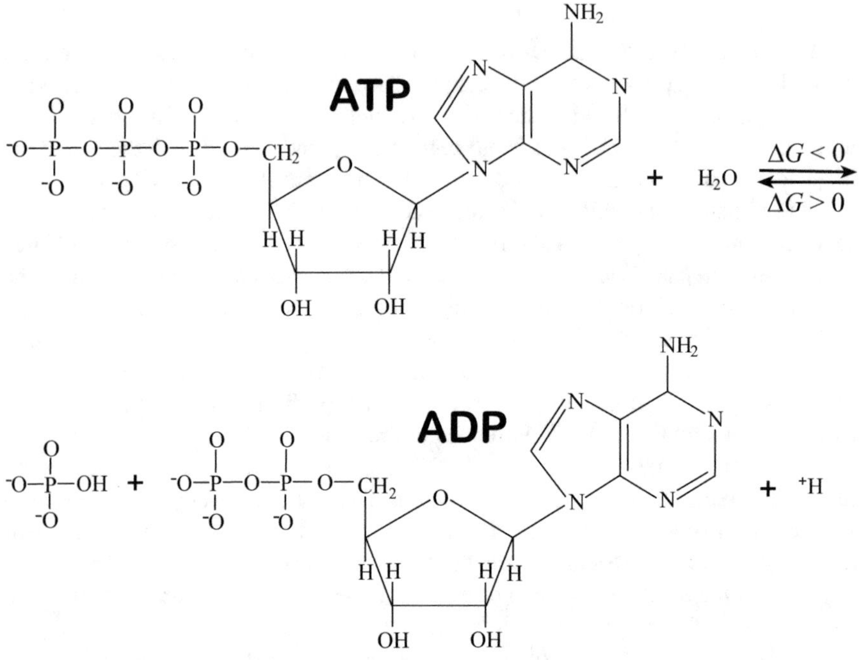

Fig. 2.7 Hydrolysis of ATP to ADP and inorganic phosphate. Molecules of ATP are the major source of energy that is required for many processes in the cells. The free energy, G, is released during hydrolysis, so $\Delta G < 0$, while the reverse process accumulates the free energy ($\Delta G > 0$). A large part of the free energy reduction comes from the formation of two molecules, ADP and phosphate, from one, ATP. Some atoms of carbon and hydrogen are not shown here

Such a reaction can proceed at a high rate due to the ability of the enzymes to couple the reactions with ATP hydrolysis. Thus, the total free energy of the substrate and ATP decreases, setting the needed direction of the reaction. Energy is also required to pump some chemical compounds through the cell membrane against their gradient (the free flow of a substance always levels its concentration through all accessible space). Of course, energy is needed for the molecular motors which perform the unidirectional movement. Pumping a substance across the cell membrane, shown in Fig. 2.8, gives an example of coupling ATP hydrolysis with a cell process that requires energy.

Let us now return to type II DNA topoisomerase (see Fig. 2.8). Although the entire reaction does not change the free energy of DNA, each of its steps can either increase or decrease the free energy of the DNA-protein complex. If the free energy change is negative, the forward direction of a step has a higher probability; if it is positive, the probability of the backward direction is higher. If we want to have a unidirectional process, each step has to be associated with a reduction of free energy. We can assume, for simplicity, that there is only one step in the reaction that increases the free energy of the DNA-protein complex. The needed forward directionality of the step is achieved by coupling this step with ATP hydrolysis. The complex makes this step due to thermal motion, regardless of ATP hydrolysis. However, in a short moment, the complex would go backward. It does not happen due to the hydrolysis of the enzyme-bound ATP and successive dissociation of ADP and inorganic phosphate from the protein. It happens very fast after the forward step is completed. This coupling greatly reduces the probability of the reverse step, which requires the binding of ADP and inorganic phosphate with the enzyme

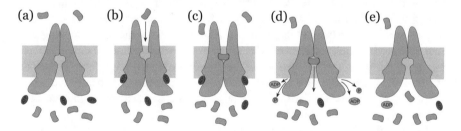

Fig. 2.8 ATP hydrolysis allows a pump to transport a substance through the cell membrane against the substance gradient. It is assumed in the figure that molecules of the substance (shown in orange) are transported down across the membrane (gray). (**a**) In the first of four successive states, the pump is closed for the substance but can bind ATP from the lower compartment. (**b**) Binding two ATP molecules triggers the allosteric transition of the pump conformation. In the new state, the pump can bind the substance. (**c**) Binding the substance causes the next allosteric transition which stimulates hydrolysis of the bound ATP molecules. (**d**) The hydrolysis and successive dissociation of ADP and phosphate from the pump decrease the free energy of the system, making this step of the process to be nearly irreversible. However, this step leaves the pump in the state with high free energy, ready for the next conformational transition. In this transition, the pump opens the exit gate for the bound substance. The transition reduces the free energy of the pump. (**e**) The association of the substance with the pump binding site is weakened in the new conformation. The substance dissociates from the pump to the lower space, returning the pump to its original state

followed by the formation of ATP from the two molecules. In this way, ATP hydrolysis enforces the needed directionality of the reaction step.

2.5   Molecular Motors

The same principle of coupling an uphill reaction step (an increase of the free energy) with ATP hydrolysis is used in biological molecular motors. These motors perform unidirectional movement and are very important for all organisms. Our muscles consist of many billions of identical molecular motors, coordinated work of which provides fast contraction of the muscles. Molecular motors perform directed translocation of molecules and their complexes inside the cell. Let us consider the major elements of a schematic molecular motor.

The majority of molecular motors are proteins or complexes of proteins that perform unidirectional movement along special protein-based filaments attached to the cell walls (Fig. 2.9). The filaments have a periodic structure with a period that corresponds to one step of the motor protein. A step of the walking protein consists of a few allosteric transitions in the protein. Thus, after the completion of one step, the protein returns to the same initial state and has the same interaction with the filament. Therefore, the protein has the same free energy before and after the step. This means that the steps forward and backward should have the same rates. The directionality of the steps is achieved by coupling each step with ATP hydrolysis. Among the allosteric transitions constituting the step, there is at least one transition that is associated with an increase in free energy. Coupling the end of this allosteric transition with ATP hydrolysis makes the step energetically favorable, introducing a very strong bias to the random directionality of the protein movement. The backward step of the protein requires the binding of ADP and phosphate to the moving protein, followed by the formation of ATP. These events would greatly increase the free energy of the system which consists of the protein, ADP, and phosphate. Therefore, the backward step has a very low probability. Correspondingly, at nearly every step the motor protein moves forward, and this movement is accompanied by energy consumption (ATP hydrolysis).

In general, the movement of motor proteins is a process of diffusion, the major type of motion in the molecular scale in water solution. The driving force of this movement is thermal motion, and ATP hydrolysis, coupled with each step of the motor, only prevents the backward steps. One can say that ATP hydrolysis rectifies the diffusion. The chemical energy of the hydrolysis does not direct or accelerate the random movement of a motor protein, it is only used to make backward movement from certain points of the protein step to be nearly impossible.

It may seem surprising that diffusion is capable of providing a very high speed of muscle contraction, from our macroscopic point of view. The quantitative estimations show, however, that diffusion-based processes can have a speed that we observe in the movement of living organisms, sufficient to explain even the very fast movement of wings of some small flies. In addition to the movement itself, over the

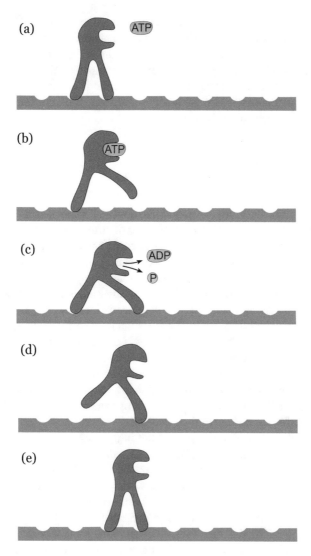

Fig. 2.9 Movement of a motor protein along the filament. The movement represents a set of allosteric transitions in the protein induced by interactions with the filament, ATP, ADP, and phosphate. Note that the filament has a periodic structure with specific sites for binding the protein "legs." In state (**a**), both legs are bound with the filament with the minimum separation between them. Transition to state (**b**) is induced by the binding of ATP. In this state, the right leg has a weaker binding with the filament and therefore is released first. The allosteric transition gives a larger separation between the walker's legs. The transition of the protein to state (**c**) is coupled with ATP hydrolysis and the release of ADP and phosphate. In this state, the affinity of the right leg to the binding site of the filament is high, so it binds the site. The binding causes a transition to state (**d**), where the left leg has a weaker interaction with the filament. Release of the left leg causes the last allosteric transition to state (**e**). The transition reduces the separation between the legs and increases the affinity of the left leg to the filament. The binding of the left leg with the next site on the filament follows the latter transition. As a result of this cycle of allosteric transitions, the walker returns to the original state but shifts to the right on one period of the filament

time of one step, the muscle cells have to receive a signal for muscle contraction. This signal is provided by a fast increase of Ca^{2+} concentration in the *cytoplasm* (the cell interior) of the muscle cells. This increase is due to the opening of Ca^{2+}-release channels, which allows the ion diffusion from a compartment with high Ca^{2+} concentration to the cytoplasm where their concentration is many times lower. So, the signal for the contraction is also provided by diffusion. Still, the entire process can occur in the time scale of milliseconds.

2.6   Energy Transfer from Food to ATP

The conversion of energy stored in food into new ATP molecules represents a long chain of chemical reactions catalyzed by dozens of different enzymes. Consideration of these reactions is beyond the scope of this book, and below, we only briefly outline the major stages of the process. A great part of the food consists of proteins, polysaccharides, and fats (compounds whose main components are fatty acids). In the first stage, these macromolecules have to be digested into corresponding monomeric units: proteins into amino acids, polysaccharides into sugars, and fats into fatty acids and glycerol (Fig. 2.10). The digestion is an energetically favorable reaction, and some energy of the macromolecules is transformed into heat at this stage. Digestion is necessary to obtain a limited assortment of substrates for other stages of the process. This first stage of conversion occurs outside the cells since it often requires an environment and enzymes which are harmful to the cell interior. The digestion in animals is performed in special compartments (stomach and intestines), while fungi perform it outside of their bodies.

Sugars, fatty acids, and amino acids enter into the cytoplasm of the cell, where the chains of their chemical transformations continue. First, all sugars are transformed into glucose. Glucose then enters into the chain of reactions called glycolysis. Glycolysis of a single molecule of glucose results in the production of two ATP molecules from two ADP and phosphates. Also, two molecules of NADH, another important energy carrier, are produced in glycolysis. Glycolysis does not require molecular oxygen (O_2), and therefore, it occurs in the same way in nearly all cells, including anaerobic microorganisms. In the latter case, glycolysis is the principal chain of reactions producing ATP and NADH. However, a much larger part of chemical energy remains in two molecules of pyruvate, the products of glycolysis. Further transformations of pyruvate for ATP production require molecular oxygen. In animal muscle cells, when they do not receive enough oxygen, glycolysis also becomes a major source of ATP. However, under normal conditions, aerobic cells produce many more ATP molecules in further transformations of two molecules of pyruvate.

In aerobic cells, molecules of pyruvate, fatty acids, and amino acids are transformed into molecules of acetyl CoA. The latter molecules enter into the next stage of reactions, called the citric acid cycle or the Krebs cycle. The major products of the cycle are CO_2, which is a waste of the process, and NADH, the carrier of

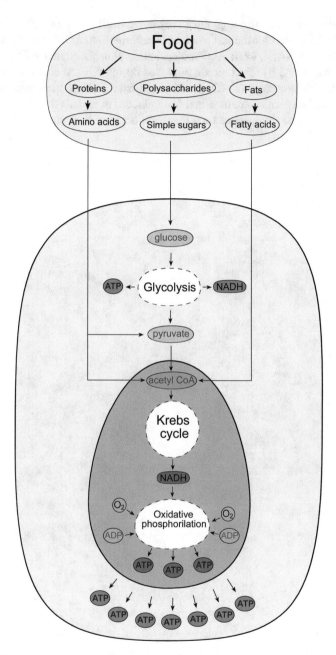

Fig. 2.10 Digestion of food in animals. The first stage of digestion, depolymerization of polysaccharides, fats, and proteins, occurs outside the cells. The obtained monomers enter the cell cytoplasm (the cell is shown in light green), where the chain of reactions, known as glycolysis, converts sugars to ATP molecules and pyruvate. Pyruvate, amino acids, and fatty acids enter into the cell mitochondria, the energy production factories located inside eukaryotic cells (a mitochondrion is diagramed in light brown). In mitochondria, all the above compounds are converted into acetyl CoA that enters into the long chain of chemical reactions. A great part of ATP molecules is synthesized in this chain from ADP and phosphate

high-energy electrons. Although the Krebs cycle itself does not require molecular oxygen, O_2 is needed for the last stage of transformation. At this last stage, NADH is converted into NAD^+, while ADP molecules are phosphorylated to form ATP. The great majority of ATP molecules are produced during this last chain of transformations, about 30 molecules of ATP are obtained from a single molecule of glucose. Only two ATP molecules are obtained from glucose in glycolysis.

Of course, some products of food digestion are used to synthesize many important biological molecules.

Chapter 3
The Cells

It was said at the beginning of this book that life on our planet goes on fundamentally in a single universal way. Maybe the most fundamental feature of this way is that it is a life of cells. It seems that although many features of life could be different from what we have on Earth, the cellular principle of life is irreplaceable. Indeed, numerous chemical processes of life require a sufficiently high concentration of substrates, enzymes, templates, and fuel needed for many reactions. On the other hand, the concentrations of waste products have to be sufficiently low. These conditions can be only created and controlled in a restricted space. The interior of the cells represents such space. Since all major features of cells are universal, to understand the basic principles of life, we should, first of all, consider an organism consisting of a single cell.

All cells have a membrane that separates the internal space of the cell from the environment. There are two major types of cells, *prokaryotic* and *eukaryotic cells*. Prokaryotic cells are simpler and include two major domains of life, bacteria and archaea, which are unicellular organisms, also called *prokaryotes*. These cells do not have internal membrane-based compartments, so each point of their internal space, the *cytoplasm*, is accessible to the cell molecules. The main distinguishing feature of eukaryotic cells is compartmentalization. These cells have many compartments for specific metabolic processes, separated by the membranes from the rest of the cell interior. Correspondingly, the content of various molecules is different in different compartments. The most important compartment is the cell's nucleus, the location of DNA. The cells of plants, animals, and fungi, including unicellular yeast, are eukaryotic. Organisms that consist of *eukaryotic cells* are called *eukaryotes*.

3.1 The Lipid Membrane

The cell membrane represents a *lipid bilayer* with numerous proteins embedded into it. The lipids, also called phospholipids, are relatively short chemical chains with hydrophilic heads and hydrophobic tails. Interaction of the membrane with

© The Author(s), under exclusive license to Springer Nature Switzerland AG 2023
A. Vologodskii, *The Basics of Molecular Biology*,
https://doi.org/10.1007/978-3-031-19404-7_3

(a)   (b)

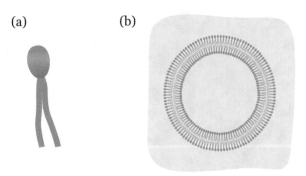

Fig. 3.1 Diagrams of a lipid and a lipid bilayer. (**a**) The lipid molecule consists of a hydrophilic head and two hydrophobic tails. The length of the molecules is about 3 nm. (**b**) When placed in water, the lipids can self-assemble into a spherical bilayer. Hydrophobic tails of the lipids contact only with each other in the bilayer, rather than with surrounding water molecules. The heads of the lipids contain hydrophilic phosphate groups, and in the bilayer they are exposed to water

water is crucial for life because water constitutes about 70% of the cell mass. Placed in water, the lipids self-assemble into bilayers, hiding their hydrophobic tails inside (Fig. 3.1). Self-assembling of complex structures is one of the major properties of biological molecules, as we will see again and again through this book.

Although all membrane lipids have hydrophobic heads and hydrophilic tails, actual membranes include a few different types of lipids. Also, the lipid compositions of the inner and the outer monolayers are different, in accordance with their different functions in cell life. In multicellular organisms, lipid compositions are different in different cells, helping these cells to perform their specific functions.

The membrane can be electrically polarized, due to some difference in the number of ions absorbed at two sides of the bilayer. As we will see later, polarization can regulate the state of transmembrane channels and plays a key role in signal transduction.

At physiological temperatures, lipid bilayers behave like two-dimensional fluids, allowing fast movements of the membrane components. Each lipid can move within its monolayer. Such movement does not change the monolayer notably, although it facilitates the incorporation of numerous proteins into the membrane.

3.2   Membrane Proteins

3.2.1   General View

Proteins incorporated into the lipid bilayer constitute about 50% of the membrane mass. The majority of these proteins stick out on both sides of the membrane; they are called transmembrane proteins. Still, some membrane proteins are only exposed to either the interior or exterior of the membrane (Fig. 3.2). The surface of

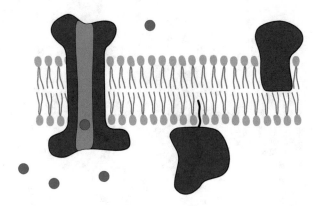

Fig. 3.2 Various types of membrane proteins. The membrane channels (on the left) represent the transmembrane proteins exposed to both sides of the membrane. The channels allow molecules (shown as small circles) to pass across the membrane. Other proteins are embedded only in a single monolayer of the membrane (on the right). Besides, proteins can be connected with the membrane by a covalently attached lipid chain (in the middle)

transmembrane proteins has a hydrophobic area interacting with the hydrophobic interior of the membrane and a hydrophilic area at the parts exposed to the cytoplasm and the cell exterior. This feature of the proteins makes embedding such proteins into the membrane energetically favorable. The embedding is greatly facilitated by the fluidity of the membrane bilayer. The transmembrane proteins are crucially important for cell life, and their variety is enormous. They constitute about 30% of all proteins coded in an organism's *genome*. The term "genome," which we will often use below, means the entire set of unique DNA molecules of the organism.

Some small molecules are capable of diffusing across the lipid bilayer, while others are not able to do so. Hydrophobic molecules and gases like oxygen and carbon dioxide cross membranes rapidly. Small polar molecules, such as water, can also diffuse through membranes, but they do it more slowly. The bilayer, however, strongly restricts the diffusion of highly charged molecules and large molecules, such as sugars, amino acids, and bases of nucleic acids. This limited permeability of the membrane allows cells to regulate the concentrations of various molecules and ions in the cytoplasm and internal compartments. The regulation is accomplished by the transport of various molecules in and out of the cytoplasm by membrane transport proteins. There are two classes of these proteins, channels and transporters, which we consider separately in the next two sections.

3.2.2   Channels

Channels allow many small molecules, first of all ions, to pass across the membrane in either direction. Each type of channel usually has high specificity for the molecules that can pass through it. Channels provide the so-called passive transport of

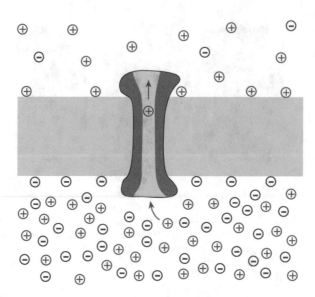

Fig. 3.3 The flow of ions through an ion channel. There is an excess of positive ions over negative ones in the upper compartment and an excess of negative ions in the lower compartment. However, there are more positive ions in the lower compartment than in the upper one. Due to this, the positive ions preferentially move upward, although the membrane electric potential stimulates their movement in the opposite direction. The majority of channels have high specificity to the ions which can pass through them. In this illustration, the negative ions cannot move through the channel

molecules so that the flow of the molecules is due to the difference in the molecule concentration outside and inside the cell (or a cell's internal compartment). In the case of ions, their flow across the channels is also influenced by the membrane's electric potential. Therefore, the direction of flow is specified by a combination of these two factors quantitatively expressed by the *electrochemical gradient* (Fig. 3.3).

The majority of channels can be in the open or closed state. The channels for water molecules, *aquaporins*, represent an exception—they are always open. Remarkably, these channels do not pass ions, including H^+. The majority of channels are ion channels, which pass Na^+, K^+, Ca^{2+}, and some other ions. In most cases, they have very high selectivity to specific ions. For example, the channels specific to K^+ do not allow Na^+ to pass through them. Since Na^+ ion is smaller than K^+, this selectivity puzzled researchers for a long time, and only establishing the 3D structure of the channels allowed explaining it. Ion channels play a very important role in many processes, including muscle contraction and signal transmission in nerve and muscle cells. This signal transmission will be considered in Chap. 7. Of course, channels have to be complemented by the corresponding transporters, making any changes in the ion environment reversible.

The channels can be closed/opened by binding specific ligands (small molecules that are capable of binding specific sites of proteins and affect their properties), by

Fig. 3.4 The structure of bacterial mechanosensitive channel MscL (Rees et al. *PDB*, 2006; PDB ID: 2OAR). This nonselective channel can provide a large flow of ions and other small molecules when it is open. It opens in response to a sudden increase in the osmotic pressure inside the cell. The increased osmotic pressure extends the membrane (shown as a light gray layer), and the channels open. Ions and other small molecules flow out of the cell through the open channels, and the osmotic pressure inside the cell decreases. The transmembrane domain of the channel is formed by ten α-helices. Slight mutual displacement of these α-helices causes the opening and closing of the channel. The image from RCSB PDB was obtained with the Molecular Biology Toolkit (Moreland et al., 2005, *BMC Bioinformatics*, 6:21)

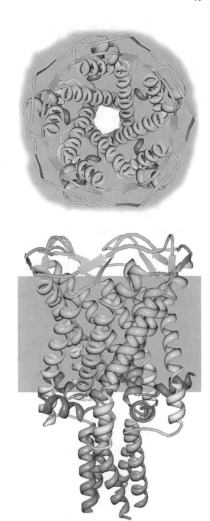

membrane voltage potential, mechanical stress, chemical modification of the protein that forms a channel, and light. An example of a channel that is opened by the pressure increase inside the bacterium cell is shown in Fig. 3.4. The internal pressure is essentially *the osmotic pressure* (see below), which depends on the difference in concentration of various molecules inside and outside the cell. Changes in these differences could strongly increase the internal pressure and cause cell explosion. Therefore, the cell has to have mechanisms to avoid large increases in osmotic pressure. The mechanosensitive channels provide one such mechanism. As an increase of the internal pressure extends the cell membrane, the channel conformation changes in response, opening its gate.

3.2.3   Osmotic Pressure

Suppose that we have a solvent in a vessel that is separated into two parts by a membrane permeable for this solvent (Fig. 3.5). Osmotic pressure appears if a solute, for which the membrane is impermeable, is added into the vessel so that its concentrations in the two parts are different. At thermodynamic equilibrium, the concentrations of the solute in both parts have to be even. The only way for the system to approach the equilibrium is by pouring the solvent through the membrane, from the part with a lower concentration of the solute into the part with a higher concentration. Such pouring increases the total solution pressure in the part with a higher concentration of solute. Eventually, this increased pressure stops the flow of the solvent from one part of the vessel to the other. The additional pressure created by the solvent flow is called the osmotic pressure. The equilibrium value of the osmotic pressure is proportional to the difference in molar concentrations of the solute in two parts of the vessel. In general, the total osmotic pressure is the algebraic sum of contributions from various solutes that cannot pass through the membrane.

The cell membrane has low permeability for the majority of solutes inside the cell, but it is relatively permeable for water. Besides, the membrane has special water-specific channels, aquaporins, which are always open. Large nucleic acids and proteins have very low molar concentrations inside the cells, so they make a negligible contribution to osmotic pressure. But there are many small solutes whose concentration is higher inside than outside the cell. These solutes can create so high

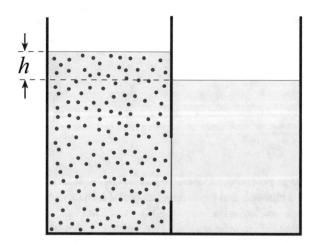

Fig. 3.5 On the nature of osmotic pressure. The vessel is divided into two parts by a wall with a membrane (shown in red/gray), which is permeable for the solvent but impermeable for the solute. If all solute is in the left part of the vessel, the solvent starts poring through the membrane from right to left, to equilibrate the solute concentrations in both parts of the vessel. The flow of the solvent stops when it creates sufficient additional pressure in the left part of the vessel. The pressure created by this mechanism is called osmotic pressure. Its magnitude is proportional to h in the drawing

osmotic pressure that it will be capable of exploding the cell. However, some solutes have many times higher concentrations outside the cell than inside, first of all Na^+ and Cl^-. They make the opposite contribution to the total osmotic pressure, which nearly compensates for its internal component.

An interesting example of fighting with a burst of osmotic pressure show some bacteria which live in water. Sometimes the concentrations of ions in water sharply decrease, due to the rain, for example. This causes a large increase in osmotic pressure inside the bacteria and swelling of their membranes. The swelling opens large emergency mechanically gated channels (see Fig. 3.4) in the membrane allowing many small molecules to diffuse out of the cell to rescue the bacteria from the explosion. When the ion concentration in the environment returns to normal, the gates of these channels close. After this, the inside concentrations of small solutes start increasing again by the corresponding transporters.

3.2.4 Transporters

Transporters can perform active transport, so the flow of the molecules can increase the difference in the concentration of the molecules outside and inside the cell. Similarly, the transport of ions can increase the transmembrane electric potential. Of course, active transport requires energy, and transporters couple their work with energy consumption. The energy can be provided by hydrolysis of ATP (see Chap. 2), absorption of light, or coupling the transfer with the transfer of other molecules in the energetically favorable direction. A transporter whose work is coupled with ATP hydrolysis was considered earlier (see Fig. 2.8). The structure of an actual transporter is shown in Fig. 3.6. Transporters can serve for passive transport which does require additional energy, although such transport through the channels is much more efficient. The latter ones can pass up to 100,000 times more ions per second than the transporters. This is understandable because passing each small molecule across a transporter requires binding of the molecule with the protein followed by structural changes in the transporter. Passing through a channel by a small molecule occurs by its diffusion without binding to the channel interior and without structural changes in the protein.

Interestingly, special membrane transporters provide cell resistance against many drugs. These transporters simply pump out certain antibiotics from the cell. This mechanism of antibiotic resistance has been found in human cancer cells, in the malaria parasite, and in bacteria. The existence of such transporters is not so surprising since antibiotic is a natural weapon in the fight for survival in the world of bacteria and other organisms. So, there was enough time for the evolution to develop such a defense system. As we discuss later in this book, the genes coding such transporters are spread between species by *parallel gene transfer*.

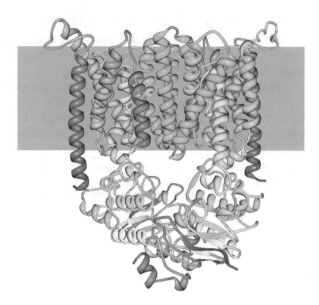

Fig. 3.6 Structure of bacterial ABC transporter involved in the uptake of B12 vitamin (Locher et al., *PDB*, 2002. PDB ID: 1L7V). The cell membrane is shown in gray. The image from RCSB PDB was obtained with the Molecular Biology Toolkit (Moreland et al., 2005, BMC Bioinformatics, 6:21)

3.3   Prokaryotic and Eukaryotic Cells

Prokaryotic cells are cells of bacteria and archaea, unicellular organisms that constitute two major domains of life. The phenotypes (observable characters) of the bacteria and the archaea are not so different. The difference between these two domains becomes clear from a comparison of their genes, as we will discuss in some detail in Chap. 4. There is no necessity for the prokaryotic cells to interact one with another, although sometimes they do it. They are very small, with a linear size in the range of a few μm. DNA of a prokaryotic cell consists of a single *chromosome*, a complex of a DNA molecule with special structural proteins. The chromosome is in direct contact with the cytoplasm. The genomes of prokaryotic cells consist of a few thousand genes. The majority of prokaryotic cells are surrounded by the cell wall, a rigid layer of polysaccharides located outside the cell membrane, which protects them from large harmful macromolecules and increases their mechanical rigidity. The biochemical variety of prokaryotic cells is enormous. In particular, they can use very different sources of energy to support themselves. For this goal, various types of prokaryotic cells can consume various types of organic compounds like sugars, amino acids, and methane gas. Some cells use hydrocarbons as food—thanks to them the oil spills in the seas eventually disappear. Others use the energy of light. Some cells use simple inorganic compounds for their energy needs. Still, beyond the treatment of energy sources, the major biochemical processes are the same in all cells. All prokaryotic and eukaryotic cells have more than 200 common families of genes. Although the sequences of these genes are not identical, their assignment to a particular family is unambiguous. A comparison of these sequences has become an important research instrument (see Chap. 4).

Eukaryotic cells are about ten times larger than prokaryotic ones. They have a complex internal organization with a set of compartments isolated by bilayer membranes from the rest of the intracellular space. The nucleus, which contains the cell DNA, is the most important of these compartments. Others are involved in the digestion, secretion, and production of fuel compounds. Compartmentalization allows the cell to have different environmental conditions for different processes. Eukaryotic cells do not have a cell wall, and as a result, their shape is more flexible. Their shape is set by a system of intracellular filaments that connect different parts of the cell membrane. The filaments crisscross one another forming an elaborated network. During cell life, the length of each filament can be changed, and this changes the entire network and the cell shape as well. The filaments are also used by motor proteins, which move along them to deliver cargo from one part of the cell to another.

Eukaryotic cells appeared from prokaryotic cells later in evolution. Many unicellular organisms eat others by engulfing them, and this process contributed to the appearance of eukaryotes. There are solid proofs that it happened in the case of the mitochondrion, an organelle responsible for aerobic (oxygen-based) ATP production in eukaryotic cells. Mitochondria have their own DNA, tRNAs, and ribosomes, which are different from those elsewhere in the eukaryotic cells. At some moment in the evolution, an ancestral cell engulfed an oxygen-metabolizing bacterium. Somehow the bacterium escaped digestion and evolved in symbiosis with the engulfing cell. Because aerobic ATP production provides a much more complete and efficient way of extracting energy from food, mitochondria spread out eventually to all eukaryotes. It happened more than 1 billion years ago.

Multicellular organisms (including ourselves) consist of a huge variety of differentiated cells, which dramatically differ one from another. However, all these cells originated from a single cell, the *zygote*. It is important to emphasize that despite their difference, all cells of a multicellular organism have the same genome. They differ only due to different expressions of various genes. Regulation of differentiation is a complex process, which will be discussed later.

All different types of cells of a eukaryotic multicellular organism can also be divided into two categories, *somatic cells* and *germ cells*. These two categories of cells are formed at the very beginning of organism development and have different functions in the life of the organism. The germ cells can produce *gametes* upon their division, which are cells that have the potential to form the zygote of a new organism after fusion with a gamete of different sex. The germ cells transmit genetic information from parents to offspring. In some sense, the germ cells can be immortal. All other cells of the organism are somatic cells. Somatic cells serve the life of the current organism. Any changes in their genes will not be transmitted to the offspring. Over the life span of the organism, they divide as well, but their division can only produce new somatic cells. Somatic cells are never converted into germ cells, and vice versa (Fig. 3.7).

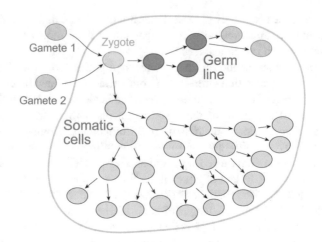

Fig. 3.7 Two types of cells of sexually reproducing organisms. Germline cells provide transferring genetic information to progeny, while somatic cells constitute the body of the organism

3.4   The Cell Cycle

The cell cycle starts when a new cell is formed by the division of the parental cell and ends with the division of this cell. All components of the cell such as membrane lipids and proteins, ribosomes, and enzymes have to be nearly doubled in numbers before the cell division. In eukaryotic cells, all cellular organelles also have to be duplicated. Finally, the cell genome has to be duplicated before the division. All these processes have to be carefully regulated and synchronized.

There are special problems with the cell genome. First, exactly one full copy of the genome has to go into each daughter cell. Even in the prokaryotic cell, whose genome consists of a single DNA molecule, a special mechanism is required for error-free completion of the task. The task is more difficult in eukaryotic cells, which have many chromosomes, and the daughter cells have to receive one copy of each. A special, carefully regulated dynamic network of microtubules, uniform tubes made of the special protein tubulin, directs the process. The network is bound with each pair of chromosomes which allows it to accomplish this task (Fig. 3.8).

The second problem of DNA partitioning in the daughter cells originates from the helical structure of the double-stranded DNA, where complementary DNA strands are entangled with one another. During DNA replication, each parental DNA strand is used as a template. Therefore, the newly synthesized double helices have to be entangled as well. Only due to special enzymes, DNA topoisomerases, which catalyze the passing of one DNA segment through another (see Fig. 2.6), the daughter DNA molecules become disentangled.

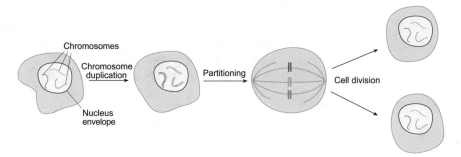

Fig. 3.8 Duplication and partitioning of the chromosomes during the division of a eukaryotic cell. At a certain moment of the cell cycle, each chromosome is duplicated. The sister chromosomes remain bound to one another. A special apparatus consisting of a network of microtubules and other proteins pull chromosomes of each pair to opposite ends of the cell. The nucleus envelope breaks down during this stage. As a result, two daughter cells receive one copy of each chromosome. At the end of the cell division, the nucleus envelope around the chromosome sets in each daughter cell is reformed

Chapter 4
Genome

4.1 Preliminary Remarks

DNA structure, discovered in 1952, immediately made clear that genetic information is written in DNA as a sequence of four letters (nucleotides). It remained, however, to be understood how the information is written there and what exactly is written. The first question was about coding the sequences of amino acids in proteins. It took about 10 years to establish that each amino acid is coded in DNA by a triplet of the nucleotides (see Sect. 1.6). The researchers understood, however, that some additional information is written in DNA sequences. Thus, it was extremely important to obtain the DNA sequences of real organisms. For more than two decades, sequencing of nucleic acids remained to be an extremely difficult task. It took years of hard work to establish sequences of a few tRNA molecules consisting of about 80 nucleotides. The breakthrough happened in the late 1970s, and in 1977, complete DNA sequence of the small virus φX174 was obtained. The genome of the virus consists of 5386 bases. This achievement marked the beginning of a new era in molecular biology. Today, genomes of thousands of organisms have been sequenced, including the human genome whose length exceeds $3 \cdot 10^9$ bp. These data brought an enormous amount of information, often absolutely unexpected, about the organization of life on our planet. Still, many pieces of information written in genomes remain to be understood. In this chapter, we will discuss what has been learned from the obtained sequences of genomes. The methods of DNA sequencing and the developed ways of DNA manipulation are quite amazing by themselves, and we will touch on this issue later in the chapter.

4.2 Prokaryotic Genome

The bacterial genome consists of a single DNA molecule. The size of these genomes can be as small as 140,000 bp (140 kb) but can reach 15,000 kb. Many hundreds of bacterial genomes have been sequenced over the last two decades.

A. Vologodskii, *The Basics of Molecular Biology*,
https://doi.org/10.1007/978-3-031-19404-7_4

N- Arg Arg Stop Asn Ala Asp Leu Ile Arg Tyr Arg Thr Ser -C
N- Stop Ala Leu Glu Cys Stop Pro Asn Gln Val Pro Asn Stop -C
N- Ile Gly Val Arg Met Leu Thr Stop Ser Gly Thr Glu Leu Val -C

```
5'-ATAGGCGTTAGAATGCTGACCTAATCAGGTACCGAACTAGTA-3'
3'-TATCCGCAATCTTACGACTGGATTAGTCCATGGCTTGATCAT-5'
```

C- Tyr Ala Ile Ser His Gln Gly Leu Stop Thr Gly Phe Stop Tyr -N
C- Leu Arg Stop Val Ala Ser Arg Ile Leu Tyr Arg Val Leu -N
C-   Pro Thr Leu Ile Ser Val Stop Asp Pro Val Ser Ser Thr -N

Fig. 4.1 Possible sequences of amino acids coded by a DNA segment. Both DNA and RNA sequences are read in the 5′ to 3′ direction. Three amino acid sequences at the top were obtained for the upper DNA strand by shifting the reading frame by one nucleotide. Similarly, three amino acid sequences at the bottom are for the lower DNA strand. Each shown segment of the polypeptide chains includes at least one stop codon (shaded as light gray). Only one segment contains the start codon, which also codes methionine (shaded as dark gray)

When the sequence of a new genome is obtained, the first task is to find DNA segments that code proteins. In the case of prokaryotes, this task is relatively simple. First of all, one needs to find all *open reading frames*, DNA stretches that begin from the start codon and end by the nearest stop codon. Since there are three stop codons, in a random DNA sequence open reading frames are relatively short, with around 20 nucleotides on average. On the other hand, the smallest protein consists of more than 60 amino acids, which correspond to 180 nucleotides. So, there is a good chance that open reading frames, which are longer than 200 nucleotides, are protein-coding genes. The search for open reading frames has to be performed for both DNA strands because, in general, each strand codes proteins. Also, in each DNA strand, the analysis has to be performed for all three reading frames (Fig. 4.1).

Another very important approach to decoding DNA sequences is based on comparing the genomes of different organisms. Since all species on Earth have a common predecessor, there is a similarity in their genomes. Of course, the DNA sequences of species evolve. It happens due to random mistakes in DNA replication during cell divisions. If they are not repaired, the mistakes cause changes in the DNA sequence, called *mutations*. Mutations also appear due to DNA damage by radiation and by various reactive chemicals, many of which are normally present inside the cells. The great majority of these damaged nucleotides are successfully repaired, as we will discuss later in Chap. 6. However, a small fraction of them is not repaired correctly. These damages are transformed by the repair machinery into normal nucleotides, but different from the original ones. Thus, DNA damage can cause inheritable genomic changes. Among these genomic changes, called mutations, only a tiny fraction is beneficial for cell survival. The majority of mutations are neutral (do not affect cell life) or harmful. The species with harmful mutations eventually die. This process is called *purifying selection*. Therefore, the mutations in DNA segments that code proteins and functional RNA molecules accumulate very slowly. As a result, proteins and RNA molecules from different species that

```
E. coli     GAAATTACTAAAAACGCGATCCGCCAGGCATTTAACAAACCGGGTGAGCTGAATATTGAT
            ||||| || ||||| || || |||||||| ||| |   | |||||| |||||||| ||| |
Salmonella  GAAATAACCAAAAATGCCATTCGCCAGGCGTTTGAACAGCCGGGCGAGCTGAACATTAAC
```

Fig. 4.2 Comparison of two segments of gene topA of bacteria *E. coli* and *Salmonella*. Vertical lines mark identical nucleotides of the genes. In 13 positions, the segments have different bases. Although in this example, the comparison shows no deletions or inserts of nucleotides in the segments, the homology search has to account for such possibility

perform the same function usually have similarities (homology) in their sequences. Of course, the homology is lower if two species diverged earlier in the evolution from a common ancestor. However, the homology maintains even for species that diverged hundreds of millions of years ago. Therefore, a comparison of protein sequences, as well as sequences of nucleic acids, has become the major tool of *evolutionary genetics*, a field of science that studies the evolutionary relations between species (see Chap. 6). A comparison of protein and DNA sequences also helps in the search for genes in the newly sequenced DNA and in establishing the protein functions (Fig. 4.2). For these goals, one has to compare the newly obtained DNA sequence with well-studied sequences where genes and their functions have been established.

The majority of genes code sequences of amino acids of proteins, so they are transcribed into mRNA or pre-mRNA (see below). There are genes, however, that code transfer and ribosomal RNAs, as well as some other small RNA molecules, participating in the regulation of various cellular processes. The nucleotide sequences of tRNAs and rRNAs are highly conserved through evolution, so the sequence comparison works even better for the identification of the corresponding DNA segments.

Other more laborious methods are also used to identify genes in the sequenced genomes. There is no universal method for this identification, however. The task is especially difficult in the case of eukaryotes, as we will discuss in the next section.

To transcribe a gene from a DNA molecule, the RNA polymerase has to recognize where on the genome to start and where to finish the synthesis of RNA molecules. Since there is a large excess of start and stop codons in DNA that do not specify any genes, the codons cannot be used for this purpose. Therefore, the transcription starts at special DNA sites called *promoters* located upstream of the genes. They are recognized by RNA polymerase that binds with them and starts the synthesis of RNA molecules. The sequences of promoters follow a certain consensus motive rather than are uniquely specified (Fig. 4.3).

The end of bacterial transcription is defined by a sequence that allows the formation of an RNA hairpin after the transcription of the segment. An example of such a hairpin is shown in Fig. 1.11.

Promoters of eukaryotes are organized similarly. They have four segments, six to seven nucleotides each, with certain sequence motives. The length of the eukaryotic promoters is about 70 nucleotides. The sequence-specific segments interact with a few proteins which assemble into a transcription initiation complex. Similar to the

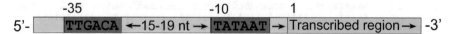

Fig. 4.3 The consensus structure of bacterial promoters. The promoter is specified by two 6-nucleotide DNA segments, which start at 35 and 10 nucleotides upstream of the transcription start. The sequences of actual promoters deviate in a few positions from what is shown in the figure. The degree of this deviation defines the promoter strength which specifies how often the RNA polymerase starts transcription from the promoter

initiation, the termination site of eukaryotic transcription is also specified by certain motives in the DNA sequence.

The expression of any gene has to be regulated according to the cell's needs at each moment of its life. Although regulation of transcription is not the only way that cells use to regulate gene expression, it is the major one. The regulation of transcription is performed by special proteins, enhancers and repressors, which bind with promoters and stimulate or suppress the transcription initiation.

4.3   Eukaryotic Genome

The size of eukaryotic genomes varies enormously, from $3 \cdot 10^7$ to 10^{11} bp. The DNA of eukaryotes is usually packed in a few chromosomes. Each chromosome consists of a single DNA molecule and many proteins bound with it. Many higher eukaryotes have a diploid set of chromosomes, so each chromosome is presented by two *alleles*, one from each parent (due to mutations in DNA, the alleles are not identical, although they code the same genes). Thus, the genome of such organisms corresponds to half of their total DNA. In general, prokaryotic genomes are very compact, and over 90% of their DNA code genes. This is very different from eukaryotic genomes. From what we know about the human genome today, it contains about ten times more genes than a typical bacterial genome, but its length is about 1000 times larger. Overall, only about 1.5% of the human genome code sequences of protein. More than 40% of the genome constitutes *transposable elements* (*transposons*), parasitic DNA distributed at different chromosomes (will be discussed below). Some of these elements have hundreds of thousands of copies in the genome, although the cell functioning does not need them at all. Each transposon consists of tens of thousands of base pairs. The other 40% of the genome constitutes nonrepetitive noncoding sequences that have no known role in cell life. We cannot exclude, however, that the meaning of some of these DNA segments will be found in the future. Established genes take about 2.5% of the genome, but a great part of this DNA does not code protein sequences, although they are transcribed into RNA molecules. Large segments of these RNAs are cut out before they become mRNA. This striking phenomenon is described in the next section.

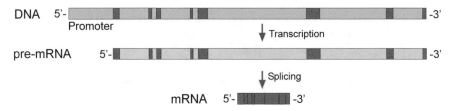

Fig. 4.4 Preparation of mRNA in eukaryotic cells. The shown gene contains eight coding segments, exons (shown in brown or dark gray), and seven noncoding segments, introns (shown in green or light gray), which are removed from transcribed pre-mRNA by the spliceosome

4.3.1   RNA Splicing

A typical gene of a eukaryotic organism consists of coding DNA segments, *exons*, and noncoding DNA segments, *introns*. The introns code RNA segments that have to be removed from the newly synthesized pre-mRNA molecules. On average, there are about ten introns in a human gene, and their total length is about ten times larger than the total length of exons. The structure and processing of a typical eukaryotic gene are illustrated in Fig. 4.4. This processing of pre-mRNA, called *splicing*, is performed by a special very large machine, the *spliceosome*, which consists of many proteins and RNA molecules.

The borders between exons and introns are specified by a few factors. The sequence of pre-mRNA makes the major input there. However, the pattern of splicing of a particular gene is set by some other factors as well. The spliceosome integrates a few signals that specify the introns that have to be removed from the pre-mRNA. Amazingly, nearly all genes of higher eukaryotes can be spliced in several alternative ways, giving different mRNA molecules. Thus, exons and introns of a particular gene are not uniquely specified. Of course, mRNAs obtained by alternative splicing produce proteins with different functions. The choice of the splicing pattern depends on the cell's needs in a particular protein at the moment. An algorithm that determines and properly integrates all signals that specify the splicing pattern is not known yet. It remains to be a challenging task to predict all patterns of splicing for a particular pre-mRNA.

The existence of introns complicates the search for protein-coding segments in newly sequenced DNA. Although the human genome was sequenced in 2003, researchers believe that more proteins are coded there than what we know now.

The phenomenon of splicing gives an evolutionary advantage to eukaryotic species, allowing them to evolve faster. Indeed, a change of a single nucleotide in the gene can change the pattern of splicing and result in the appearance of a completely new polypeptide chain. So, due to the splicing, large changes in a protein structure will be tested more often, and, eventually, a mutation that is beneficial for the organism can be selected. Without splicing, such large changes in proteins would require many single-nucleotide mutations, which will be, probably, eliminated by the purifying selection.

Fig. 4.5 Diagram of the site-specific recombination. (**a**) Two participating DNA segments are cut, and their ends are rejoined with the ends of the other segment by a special enzyme, site-specific recombinase (shown as a grey circle). The transposon ends are shown in red. (**b**) Site-specific recombination results in the insertion of the transposon into another DNA, while the reverse reaction causes transposon deletion

4.3.2 Mobile Genetic Elements

Mobile genetic elements, or transposons, represent DNA segments that are capable of moving along genomes. They have a length from a few hundred to tens of thousands of base pairs and are present in organisms of all domains of life. The movement of the transposons is assisted by special enzymes, site-specific recombinases, which interact with special DNA sequences at the transposon ends. The enzymes are capable of cutting transposons from DNA where they are located as well as inserting them into recipient DNA (Fig. 4.5). Usually, there is no sequence requirement in the recipient DNA, so the transposon can be incorporated into any location of the genome.

There are different types of transposons. Some of them move from one location in the genome to another location, without a change in their total number. They are excised from the DNA at one place and inserted at another location. Other transposons serve as templates for the synthesis of their copies, which are inserted in various locations of the genome. In the latter case, the number of transposon copies in the genome increases.

Transposons can carry a few genes. In particular, they can carry genes responsible for the resistance of bacteria to various antibiotics. Although transposons are capable of moving only within the same cell, they can be transferred from one cell to another by *horizontal gene transfer*. There are different mechanisms of horizontal gene transfer but, in general, it is a rare occasional transfer of a DNA segment from one organism to another. Of course, if a DNA segment is a mobile element isolated from the chromosome, it has a better chance for horizontal transfer. Horizontal gene transfer has an enormous impact on the development of bacterial genomes.

Similar to splicing, transposable elements can accelerate the evolution of species by randomly moving large segments of the genome. Such movement may duplicate genes. One copy of a duplicated gene can evolve much faster since mutations in this copy are not, usually, harmful to the cell. Occasionally the evolved copy may turn out to be useful to perform a new function. We know now that many genes originated from other duplicated genes. Although transposons can greatly increase the length of the genome, such excessive length is not a hard burden for large cells of higher eukaryotes. They could afford to have huge genomes where a great part of DNA, including transposons, is not needed for them.

Cells cannot survive the frequent movement of transposons since it would introduce too many changes in the genome. Therefore, eukaryotic cells developed special mechanisms to suppress the transposon movement. Also, transposons accumulate mutations that eventually inactivate their ability to move, and today, the majority of transposons in the human genome are not capable of moving. Still, some transposons in humans perform a translocation over every 100–200 generations.

4.4 Chromosomes

4.4.1 General Features of Eukaryotic Chromosomes

The total length of DNA can exceed the size of the cell by million times, and its housekeeping represents a very difficult task. This is why each DNA molecule inside the cells forms a complex with structural proteins, which compacts it enormously. These complexes are called chromosomes. Proteins of chromosomes not only compact DNA but also structuralize the molecules. These structures are very different in prokaryotic and eukaryotic cells, and many of their details have not been completely understood. Although bacterial DNA is bound with many specific proteins, it seems that it does not have a well-defined structure inside the cell. Much more is known about the chromosome structure and its functions in eukaryotic cells, which are discussed below in this section.

While the prokaryotic genome is contained in a single DNA molecule, eukaryotic cells have at least a few chromosomes. The majority of eukaryotes, including all animals, have a diploid set of chromosomes in their cells. In each pair of chromosomes, one came from the mother and one from the father. In particular, human cells have 46 chromosomes, 23 from each parent. Only one pair in the diploid set can have mother- or father-specific features. The male chromosome, Y, forms a pair with the female chromosome, X, in male somatic cells, so each somatic cell carries the XY pair. All female somatic cells have an XX pair of chromosomes. During the division of somatic cells, each chromosome is duplicated to form the *sister chromatids,* which are segregated into two new cells. Haploid gametes, which are formed by two successive divisions of special germ cells, receive one chromosome from each pair of the diploid set. Thus, male gametes carry either a Y or X chromosome, while female gametes always carry an X chromosome. A new daughter cell, a zygote, forms by fusing male and female gametes. Since it obtains all chromosomes of both male and female gametes, the zygote can have either an XY or XX pair of sex chromosomes.

X and Y chromosomes are very different in terms of the genes they code. Chromosomes of all other pairs, alleles, are very similar; they code the same proteins and RNA molecules. It is important, however, that alleles are not identical, since they have different mutations. Due to the mutations, humans differ one from another. Since the haploid set of chromosomes in a human gamete is formed by a random separation of 23 pairs of chromosomes into two sets, a huge amount of different haploid sets of chromosomes can be formed, $2^{23} \approx 8 \cdot 10^6$. It gives 2^{46} variants of the diploid sets of a human zygote. This explains why children of the same parents look different from one another.

4.4.2   Nucleosomes and Epigenetic Inheritance

The structure of the eukaryotic chromosome has a few levels of organization. At the first level, segments of DNA are wrapped around so-called *nucleosome cores*, particles consisting of eight positively charged proteins. The complex of the DNA segment with the core is called a *nucleosome*. Each nucleosome includes a DNA segment of about 150 bp. When *chromatin*, the molecule of DNA (or maybe a few of them) bound with chromosomal proteins, is gently isolated from a cell, a string of nucleosomes connected by DNA linkers can be seen in the electron microscope (Fig. 4.6). The structure of nucleosomes is nearly identical in all eukaryotes, from yeast to humans.

DNA is negatively charged (see Fig. 1.7) and therefore is strongly bound with the positively charged nucleosome core. Also, the structure is stabilized by numerous hydrogen bonds between the DNA segment and the proteins of the nucleosome core (*histones*). Although the structure of the nucleosome is subject to thermal fluctuations that release the ends of a DNA segment from the core, nucleosomes strongly repress DNA transcription and replication. To overcome the repression, eukaryotic cells have many special nucleosome remodeling complexes that modify the nucleosome structure, shift the core along DNA, or even completely remove it. The remodeling complexes are very large protein complexes that can interact with a few nucleosomes simultaneously. The majority of these remodeling complexes consume the energy of ATP hydrolysis during their operation. In general, nucleosome remodeling can make DNA more accessible for various proteins or can repress the

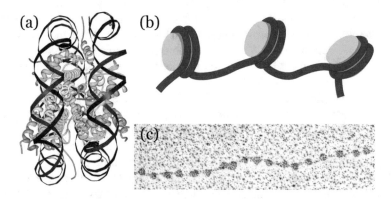

Fig. 4.6 Nucleosomes. (**a**) Crystal structure of isolated nucleosome (Vasudevan et al., *J. Mol. Biol.*, 2010. **403**, 1–10; PDB ID: 3LZ0). A DNA segment is shown in dark colors, and the protein core is in light colors. DNA makes nearly two full turns of the left-handed helix around the nucleosome core. The diameter of the nucleosome is close to 10 nm. (**b**) A diagram of three nucleosomes connected by DNA linkers. The linkers correspond to DNA segments of about 50 bp in length, on average. (**c**) Electron micrograph of nucleosomes connected by DNA linkers. The image in panel (**a**) from RCSB PDB was obtained with the Molecular Biology Toolkit (Moreland et al., 2005, BMC Bioinformatics, 6:21)

transcription of particular genes by hiding them from the proteins participating in the transcription.

How are specific sites of the chromatin chosen for a particular remodeling? The initial signal comes from the proteins which regulate transcription by binding to specific promoters. These proteins represent the major instrument of transcription regulation in the cells. The transcription regulators, bound with their specific binding sites, recruit enzymes that perform methylation of cytosines in the genes which have to be repressed. This methylation of DNA serves as a guide for chemical modification of the core histones: phosphorylation, *acetylation* (adding of an acetyl group to an amino acid of the protein), or/and methylation. There are many patterns of the chemical modification of the cores. These modifications provide various signals to the chromatin remodeling complexes, so each remodeling occurs at a particular place and at a specific time, according to the cell's needs. A particular type of remodeling involves extended segments of the chromosome consisting of many nucleosomes. Nucleosome remodeling represents a long-lasting regulation of gene expression, although the remodeling is reversible. Special enzymes can invert the chemical modifications, and the remodeling complexes can reverse the structural changes of the corresponding segments of chromosomes. However, the structural changes in chromosomes are quite stable and even inheritable. The cells have special, very large enzymatic complexes that reproduce the chemical modifications of the nucleosomes in daughter chromosomes during DNA replication. This is one of the major reasons that specific cells of the organism produce cells of the same type upon their division. This kind of inheritance of a cell's properties does not involve any changes in DNA sequence and is called *epigenetic inheritance*. Epigenetic inheritance is only related to the inheritance of the expression of specific genes. It maintains the type of specific somatic cells during their division. It should be emphasized, however, that the inheritance of any changes in somatic cells is restricted by the life span of the organism. It is also important that epigenetic changes are reversible, so they can fade away after a few cell divisions.

The expression of genes in specific somatic cells can change over the life span of the organism. Such changes can occur as a response to surrounding conditions, viral or bacterial attacks, or random deviations in the regulation mechanisms. These epigenetic changes can be inheritable over many divisions of the cells. An important question here is if these changes in gene expression can be transferred to the progeny of the organism. If this happens, it means that some features of the organism's life experience can be inheritable. For this to be possible, epigenetic changes must occur in the line of germ cells. In principle, the mechanisms of epigenetic inheritance can work in cells of the germline as well and, correspondingly, can provide this transgenerational inheritance. However, the great majority of specific chromatin marks, responsible for gene expression, are eliminated during formation of the zygote from two gametes of opposite sexes. Still, a small fraction of these marks remains in the zygote, making transgenerational inheritance possible. The study of transgenerational epigenetic inheritance in mammalians and, in particular, humans has become a hot topic over the last couple of decades. It seems that some examples of such inheritance have been found, although more data on the topic are needed.

(a)   (b)   (c)   (d)   (e)

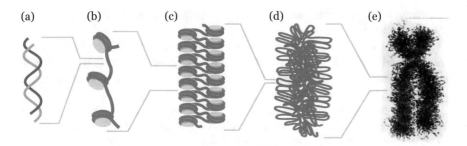

Fig. 4.7 Successive levels of chromatin packing. Panel (**a**) shows naked double helix, panel (**b**) diagrams a string of nucleosomes connected by DNA linkers. The structures shown in panels (**c**) and (**d**) are known only approximately. A micrograph of two bound identical sister chromosomes, bound together before the cell division, is shown in panel (**e**)

4.4.3   Higher Orders of Chromatin Structure

Nucleosomes represent only the first level of DNA organization in the chromosomes. At the second level, the nucleosomes are stacked on top of each other and form a helical structure of a higher order. Although the precise structure of this helix is not known, some electron microscopy data show a nucleosome fiber of 30 nm in diameter. Even less is known about the next levels of DNA folding in the chromosomes. A model of successive levels of chromatin packing is diagramed in Fig. 4.7. The chromatin remodeling involves the second and higher levels of chromosome structure.

The degree of DNA compression in the chromosomes depends on the stage of the cell cycle. The compression becomes especially high before the cell division when the duplicated chromosomes are still bound together. In this state, chromosomes look like very thick extended particles (see Fig. 4.7e). This compactness of the chromosomes facilitates their separation during cell division. The compression of chromatin is much lower for all other stages of the cell cycle. Also, it is not uniform, and its pattern is different in different types of cells. The compactness of chromatin is lower for DNA segments that contain the genes which are expressed in a particular type of cell.

4.5   DNA Sequencing

Although everybody understood the crucial importance of DNA sequencing, there was no practical solution to the problem until the second half of the 1970s. At that time, two groups developed two different approaches to DNA sequencing, which revolutionized biology. The methods allow fast sequencing of DNA segments of a few hundred nucleotides in length. One of the approaches, described below, was developed by Frederick Sanger and his colleagues. Earlier, Sanger developed a method of sequencing proteins, and for these achievements, he eventually received two Nobel prizes.

The base of both approaches was the separation of DNA molecules according to their length by gel electrophoresis, the most important and widely used physical technique in DNA-related studies. In this procedure of DNA sequencing, a mixture of DNA fragments of different lengths is loaded in the well located at the top of the gel. The gel is placed into the camera with an electrophoresis buffer where the electric field moves the negatively charged DNA in the direction parallel to the gel surface. DNA molecules of different sizes have different mobilities in the gels, so after running the electrophoresis for a certain time, typically a few hours, molecules of different sizes have different positions in the gel. The method allows the separation of single-stranded DNA molecules of different lengths with a remarkable resolution.

In Sanger's chain-termination method, a mixture of DNA fragments is synthesized on a single-stranded DNA template, in which the sequence has to be determined (Fig. 4.8). The synthesis, catalyzed by DNA polymerase, starts from a DNA primer with complementarity to a 3'-end of the template. Step by step, the enzyme extends the growing strand by attaching complementary nucleotides (dATP, dGTP, dCTP, and dTTP) to its 3'-end. In the reaction mixture, in addition to four normal nucleotides, there are four modified nucleotides. Although the incorporation of the latter nucleotides into the growing strand follows the usual rules of complementarity, it terminates the strand elongation. Thus, each synthesized fragment has a modified nucleotide at its 3'-end. Each modified nucleotide carries an attached fluorescent label of a different color for its identification.

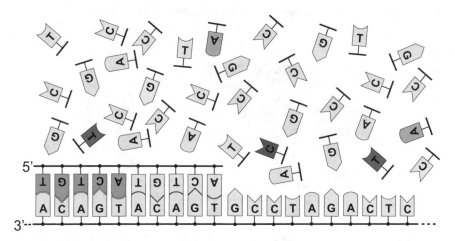

Fig. 4.8 Synthesis of DNA fragments in Sanger's sequencing method. The sequence of the single-stranded template (lower strand) has to be determined. The synthesis starts from a DNA primer, a short oligonucleotide that is complementary to a DNA segment (dark gray). Together with regular nucleotides, small fractions of fluorescently labeled nucleotides are present in the reaction mixture. The labels of four different colors are used for each of the four modified nucleotides. The incorporation of a labeled nucleotide terminates the synthesis of the upper strand. The DNA polymerase is not shown here for the simplicity of the figure

The concentration of modified nucleotides should be 2–3 orders lower than the concentration of normal nucleotides. Therefore, the average length of the synthesized fragments would be in the range of a few hundred nucleotides. The obtained DNA fragments are separated according to their size, using gel electrophoresis. Since all synthesized fragments start at the same position (from the primer), there is a direct correspondence between the fragment length and the labeled nucleotide at the 3'-end. Therefore, the fluorescent color of each band in the gel is determined by the last nucleotide at the 3'-end. The sequence of bands in the gel and their colors directly correspond to the sequence of the synthesized strand (Fig. 4.9).

Sanger's method allows the sequencing of DNA segments of a few hundred nucleotides in length. To sequence the whole genome of the organism, its DNA has to be first cut into fragments a few thousand base pairs in length. Then, after sequencing the fragments, the obtained reads have to be arranged in the proper order. There are different approaches to this problem. The sequencing of various sets of fragments followed by their computer-assisted assembling into the entire chromosome became the most popular approach now. This approach, called *shotgun assembling*, is illustrated in Fig. 4.10.

For about 30 years, Sanger's method was the main method of DNA sequencing. It allowed obtaining DNA sequences of many viruses, bacteria, and eukaryotes. It was used in the human genome project launched in 1990. The completion of this project, about 15 years later, gave the first reference sequence of the human genome. It took an enormous amount of work from many researchers, and the total cost of the project was about three billion dollars, nearly the same as the number of nucleotides in the genome. At the same time, the human genome was sequenced by a private company that used the shotgun assembly, introduced by J. C. Venter.

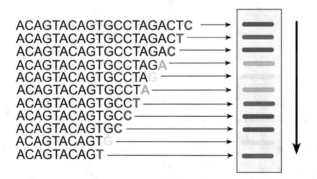

Fig. 4.9 A diagram of the DNA sequencing gel obtained by Sanger's method. The DNA fragments of different lengths are separated according to their length, so they are seen as narrow bands of different colors on the gel (shown on the right). Since the sequences of all fragments start at the same point, their length is directly related to the type of terminating nucleotide. The sequences of the fragments, corresponding to each band, are shown on the left. Each of the four colors of the bands corresponds to one of four terminating nucleotides, allowing the direct reading of the sequence. The black arrow shows the direction of the electrophoresis. Only a short segment of the gel is shown here

```
1  AGTCTGGCCATGTCG
2  AGAGCATATGCAGAGCTCC
3  ATTGCAGCGTTACAGGCTGACACGTTT

1  GGCTGACACGTTTAGAG
2  ATTGCAGCGTTACA
3  CATATGCAGAGCTCCAGTCTGGCCATGTCG
```

Fig. 4.10 The principle of the shotgun assembling of the sequenced DNA fragments. In this very simple example, two sets of sequenced fragments, dark gray and light gray, were obtained by different cutting of the same DNA molecule (shown at the bottom). Although the fragments of each set may be placed in six different arrangements, only one arrangement gives the same sequence for the whole DNA molecule. The arrows show the ends of fragments of the two sets. Finding the correct order of the sequenced fragments in more realistic cases requires extensive computation

Genomes of all humans are slightly different, and a comparison of their sequences has become an extremely powerful research instrument. At the time when the genome project was finished, the sequencing of its variants did not look practical due to a very high cost. Since then, the situation has changed dramatically, however. New, so-called "next-generation" methods of sequencing have been developed (and are under development now). More than a dozen companies now produce fully automatic sequencing machines that perform thousands of sequencing simultaneously. The cost of sequencing has been reduced by about 100,000 times. This has opened enormous possibilities for biology, medicine, and studies of evolution and human history. In particular, it opened numerous possibilities to identify variations of the genome responsible for various properties of individuals. Among them are, in particular, properties related to many diseases (see Sect. 6.4.3).

4.6   Genome Editing

4.6.1   Plasmids and Genetic Engineering

Genome editing is a laboratory procedure that changes the DNA of an organism. In its simplest form, it is the insertion of a DNA segment into the DNA of an organism. Such editing started in the 1970s, after the discovery of *restriction nucleases*, the enzymes which can introduce sequence-specific double-stranded breaks in DNA molecules. The enzymes represent a part of the bacterial defense system against the invasion of viruses (see Chap. 9). The enzymes allow the insertion of a DNA segment into another DNA molecule. The procedure is applied to small DNA circles, called *plasmids*, which can exist in bacteria in parallel with their major DNA. Plasmids consist of a few thousand base pairs and represent a supplement to the bacterial

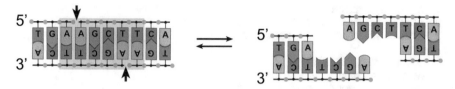

Fig. 4.11 A restriction nuclease introduces sequence-specific breaks in the double helix. The enzyme recognition site consists of six shown nucleotides (shaded segment). The single-stranded breaks (shown by arrows), introduced by the enzyme, are shifted one relative to the other by four base pairs. Although the new DNA ends can be held together by the four base pairs, at slightly elevated temperatures they can be either joined or separated from each other

Fig. 4.12 Inserting a DNA fragment into a plasmid. The plasmid has a recognition site for a restriction nuclease that binds the site and cuts both strands. The obtained linear molecules are mixed with a DNA fragment made for insertion into the plasmid. The efficiency of the insertion increases greatly if both the linearized plasmid and the fragment have the same sticky ends. Treating the mixture with DNA ligase reseals the single-stranded breaks and creates, with some probability, DNA circles with the desired insert

genome. Bacteria can have many copies of a particular plasmid that can carry genes useful for their survival. The insertion of a new segment into the plasmid is performed in a laboratory tube where only DNA and the needed enzymes are present. The plasmids, used for this simplest type of genome editing, have to have a single site with the sequence recognized by a restriction nuclease. Usually, the recognition site consists of 4–6 nucleotides. The molecules of the needed restriction nuclease are added to the solution of plasmids, bind the corresponding sites, and introduce a double-stranded break there (Fig. 4.11). After this, the molecules of the nuclease have to be inactivated. Very often, the breaks form so-called sticky ends, which fluctuate between joined and separated states. If another segment of DNA with the same sticky ends is added to the solution, there is a chance that it will be incorporated between the newly formed ends of the plasmid (Fig. 4.12). Four single-stranded breaks in the plasmid with the incorporated insert can be resealed by a special enzyme, *DNA ligase*. This step converts the plasmid back into circular form with intact complementary strands.

Although a relatively small fraction of resealed DNA forms the desired construct, it can be enough for further steps. The new DNA circles can be placed back into bacteria, simply by mixing the circles with a bacterium culture. Again, the efficiency of this procedure is not so high, but there is a way to select the bacteria which received the plasmid. For this goal, researchers use plasmids that have a gene of resistance to an antibiotic, so in the media with this antibiotic, only bacteria that have the plasmid inside are capable of growing and dividing. Extraction of the modified plasmids from these bacteria and simple tests, or even direct sequencing, is used to check whether the plasmids carry the insert. There are ways to enhance the efficiency of the entire procedure, which is often referred to as *genetic engineering*.

The insert can carry gene coding a protein and a promoter needed for the gene expression in the bacteria. If the promoter is very strong, it forces the overproduction of the desired protein by the bacteria. This way of obtaining proteins is widely used in molecular biology and biotechnology.

4.6.2   CRISPR-Cas9 and Editing of Large Genomes

The antibiotic used in the procedure described above always has to be in the media where the bacteria grow. Without it, the plasmid will be soon lost by the bacteria since they do not need the new protein and its production gives them a selective disadvantage in the absence of the antibiotic. Plasmids are not a part of the bacterial genome, so the described procedure cannot be considered as editing of the bacterial genome. To make the genetic alterations more stable, one has to change the bacterial genome itself rather than the plasmid. The first obstacle there is that the bacterial genome has too many recognition sites for the restriction nucleases, so bacterial DNA will be digested into many fragments if treated by the enzymes. To obtain a single cut in a bacterial genome, one needs a *nuclease* whose sequence-specified binding site is sufficiently long to be unique in the entire genome. It is also highly desirable that the recognizable DNA sequence can be changed according to specific needs. The search for such an enzyme took many years, but eventually, a few solutions to this problem have been found. The most efficient of them is based on another bacterial defense system against viral attacks, the CRISPR-Cas9 system (see Chap. 9). In this case, the recognition site consists of 20 nucleotides, and its sequence is specified by an RNA oligonucleotide bound with the protein that cleaves the recognition site (Fig. 4.13). The recognition is based on the formation of the RNA-DNA double helix with the complementary base pairs, AU and GC, which is similar to the DNA double helix. Since it is easy to make an RNA molecule with the desired sequence, researchers can prepare the CRISPR-Cas9 complex, which binds nearly at any DNA site and introduces the single cut in a very large genome.

The introduction of a single, sequence-specific break in the double helix is only the first step in genome editing. Editing may include the insertion or deletion of a DNA segment, and, eventually, the newly formed ends of the double-stranded DNA have to be properly rejoined. Although the cell DNA repair system can perform the latter task, rejoining DNA ends by this system often results in a loss of a few base pairs at the junction. In the case of multicellular organisms, the editing system has to be delivered into many cells. This represents another problem. Also, there is a danger that the editing system will introduce undesired harmful changes in

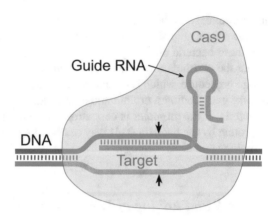

Fig. 4.13 DNA cleavage by CRISPR-Cas9. The protein Cas9 forms a complex with the guide RNA. With the assistance of Cas9, a segment of the guide RNA can form DNA-RNA double helix with the target segment of DNA, if its sequence is complementary to the segment of the guide RNA. The specific binding of CRISPR-Cas9 is followed by the cleavage of both strands of the double helix at a certain position (marked by arrows)

off-target sites of the genome. Due to these problems, the probability of successful genome editing in the cells of vertebrates remains relatively low.

Despite all these difficulties, applications of genome editing are developing at an amazing pace. One field of its application is cancer therapy. In this case, the editing can be performed on isolated cells of the patient's immune system, which greatly simplifies the strategies. So far, successful applications of the therapy are limited, however.

Another example of genome editing, which is also under development, is the therapy of the gene responsible for a muscle disorder. The disorder is usually caused by a single base replacement in a single gene. The gene is so large, however, that it does not seem possible to replace the entire mutated gene with its healthy copy in the muscle cells. Therefore, the strategies which researchers are trying to implement are directed to simpler editing, which includes a short segment of the gene. Although the majority of these editing strategies give only partially functional protein, even such protein greatly improves muscle functioning, as the experiments on animal models show.

A simpler editing procedure has been developed recently to correct nucleotide replacements. There are enzymes, both natural and humanmade, which catalyze the substitution of one specific base of the double helix by another one. Combining these enzymes with CRISPR-Cas9, researchers created CRISPR base-editing systems that catalyze base substitution at a chosen nucleotide without cutting the double helix. The data show that this system provides better results for the base substitution in the targeted position of a gene.

The majority of genome editing applications address DNA changes in somatic cells. This means that the progeny will not inherit any changes in the genomes. To make inheritable changes in the adult organism, one has to introduce changes into the germ cells, in which division produces male and female gametes. It is easier to edit embryos or even zygotes, and such attempts are underway now. Of course, in this case, the editing is associated with additional problems. Due to the danger of introducing undesired harmful changes in off-target sites of the genome, the editing cannot be applied to human embryos in the current medical practice. Although this danger slows down the experiments, it is hardly possible to stop them. The first results show that the procedures, even when they are successful, introduce unintended edits outside the targeted gene. It may take time to overcome the problem.

Overall, there are many problems in human health, which can be addressed with genome therapy, and many labs around the world are designing the needed methodologies. Genome editing is developing at a stunning pace and its application to cure gene-related diseases in humans is becoming a reality.

The danger of undesired changes in the genome is tolerable in the case of animals and is not an obstacle to editing the genomes of plants. The results of such editing, genetically modified vegetables, are widely used in our everyday life. Although such vegetables run across public opposition in many countries, there is no serious reason to think that they carry any danger. Indeed, various genome editing takes place in nature over billions of years, since the very beginning of evolution, and so far, nothing horrible has happened. There is one visible problem there, however. The most common kind of genetic modification of plants is adding to their genomes the resistance to an herbicide or pesticide. Such modification stimulates farms to apply larger amounts of chemicals, which is harmful. Besides, the weeds or pests can capture the resistance gene by *the parallel gene transfer*, so they will become resistant to the herbicide or pesticide. In this case, the modification of the plant genome becomes useless. Some genetically modified plants produce proteins that kill specific pests. These proteins are harmless for humans, so this kind of genetic modification is safer for our health.

Chapter 5
Regulation of Gene Expression and Protein Activity

5.1 Regulation of Transcription

The expression of a protein-coding gene consists of many steps and can be regulated at each of them. However, the most important regulation occurs at the level of transcription initiation. Transcription of a particular gene starts when RNA polymerase binds the promoter of the gene and starts RNA synthesis. Correspondingly, if the promoter is bound with another protein, this can block the polymerase binding. The work of the transcription *repressors* is based on this simple principle. Another class of proteins, the transcription *activators*, helps RNA polymerase to initiate transcription. Activators interact with RNA polymerase directly when both proteins are bound with DNA. The binding site for the activator is adjacent to the binding site for RNA polymerase. Of course, the affinity of these regulators to DNA has to be sequence-specific, that is, they have to have a high affinity to the specific DNA site and relatively low affinity to all other sites. Therefore, both repressors and activators have to recognize specific DNA sequences. This recognition is provided by the complementarity of interfaces between specific DNA sites and the sites of regulators that interact with DNA. How is this complementarity achieved? There was hope before the first structures of DNA-protein complexes were resolved, that there is a correspondence between DNA bases and amino acid residues in the recognition interfaces. This hope evaporated soon, however. Many different amino acid residues can interact with any of four different bases, allowing a protein to have a recognition site for a DNA segment with any sequence. The hydrogen bonds between the proteins and DNA are the major elements of recognition. The search for a specific DNA site among millions of others is facilitated by the fact that proteins can perform this search without opening the base pairs. Certain chemical groups of the bases are exposed to the grooves of the double helix, and this is sufficient for the unambiguous reading of DNA sequence by the proteins.

© The Author(s), under exclusive license to Springer Nature Switzerland AG 2023   67
A. Vologodskii, *The Basics of Molecular Biology*,
https://doi.org/10.1007/978-3-031-19404-7_5

5.1.1   Transcription Regulation in Prokaryotes

Regulation of transcription initiation is much simpler in prokaryotes than in eukaryotes. Therefore, to illustrate the operation of transcription regulators, we first consider two simple well-studied examples from bacteria. The first example is the tryptophan *operon*, a cluster of five genes under the single promoter that codes enzymes needed for tryptophan synthesis (Fig. 5.1). These genes are transcribed together to produce a single mRNA molecule. The transcription is regulated by a tryptophan repressor which can be in two conformations, active and inactive. The repressor is in its active conformation if it is bound with tryptophan; in this conformation, it can bind the *operator* (a region in the promoter which can bind transcription initiation proteins or transcription repressors) and turn off transcription of the genes. This happens when the concentration of tryptophan in the bacterium is sufficiently high, so it does not need to synthesize tryptophan (the bacterium can receive enough tryptophan from the meal). However, sometimes, the bacterium

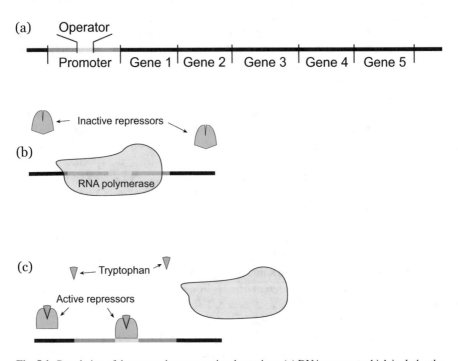

Fig. 5.1 Regulation of the tryptophan operon in a bacterium. (**a**) DNA segment which includes the tryptophan operon. The operon consists of a promoter and five genes that code the enzymes catalyzing tryptophan synthesis. (**b**) When the concentration of tryptophan in the bacterium is low, it needs the corresponding mRNA molecules transcribed from the operon. The tryptophan repressor is in its inactive form, so it does not prevent the transcription of the genes. (**c**) If the concentration of tryptophan is sufficiently high, it binds with the tryptophan repressors. The repressor changes its conformation to an active one and binds the operator located in the middle of the operon promoter. The bound repressor blocks the binding of RNA polymerase, so the operon is switched off

does not receive enough tryptophan from the meal and has to produce it. Thus, the operon has to be turned on by inactivating the repressor. The molecules of tryptophan dissociate from the repressor if the concentration of tryptophan becomes low. This switches the repressor into inactive conformation and causes its dissociation from the operator. The operon turns on, and the bacterium starts production of the corresponding mRNA and enzymes, which catalyze tryptophan synthesis.

The second example describes a system where activation and inactivation of the operon depend on two signals rather than one, as it is in the case of the *Lac* operon. The system includes both a repressor and an activator. Transcription activators work on relatively weak promoters and are capable of accelerating transcription initiation by up to 1000 times. Similar to repressors, activators can bind other molecules, and the binding switches them from active conformation to inactive, or vice versa.

The genes coded by the *Lac* operon are needed to digest the disaccharide lactose. The preferred meal for the bacterium is glucose, and if there is enough glucose in the media where the bacterium is growing, there is no need to digest lactose. Correspondingly, there is no need for the genes of the *Lac* operon. If there is not enough glucose, lactose can be used as a meal, if it is present in the media. Thus, only if there is no glucose, but there is lactose, the operon has to be turned on. The switch is achieved by the combined action of the *Lac* repressor and the activator.

The repressor can be in an active or inactive conformation. In the absence of glucose, the bacterium produces a compound called cyclic AMP, which binds the activator and converts it into an active conformation. The repressor, on the other hand, is converted into an inactive form upon binding with lactose. Four states of the *Lac* operon are shown in Fig. 5.2.

It is important to emphasize that both transcription repressors and activators act only when they are bound with their specific binding sites on DNA. Thus, the action of the regulators extends to only one or a few specific genes.

5.1.2   Eukaryotic Transcription

In bacteria, the binding sites of the regulatory proteins are located, usually, near the transcription start. This facilitates their interaction with RNA polymerase. However, in eukaryotes, the binding sites for the regulatory proteins are nearly always separated from the promoter and can be located thousands of base pairs away from it. In such cases, the regulator has to bind both the remote binding site and RNA polymerase. Thus, the DNA segment between the binding site of the regulator and the promoter is looped out to allow interaction of the RNA polymerase and the DNA-bound regulator (Fig. 5.3). It may look like binding the regulator with the remote site on DNA is not necessary at all. However, the binding has two consequences. First, it greatly increases the effective concentration of regulators in the vicinity of the promoter. Second, it may cause conformational changes in the regulator which increases its affinity to the RNA polymerase.

Simple regulation of the transcription initiation illustrated above for prokaryotes is rare in eukaryotes. At a typical eukaryotic promoter, the RNA polymerase

(a)   **No glucose, no lactose - operon is off**

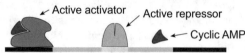

(b)   **No glucose, there is lactose - operon is on**

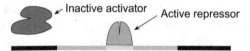

(c)   **Enough glucose, no lactose - operon is off**

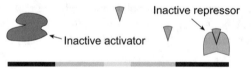

(d)   **Enough glucose and lactose - operon is off**

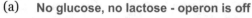

Fig. 5.2 Regulation of the *Lac* operon. The genes of the operon code proteins that are responsible for the digestion of lactose, the milk sugar. The operon should be on only when the amount of glucose in a bacterium meal is insufficient, and there is lactose in the media. This is achieved by the work of Lac repressor and the activator of the operon. When there is no glucose in the meal, the bacterium produces cyclic AMP that binds the activator and converts it into its active conformation. In this conformation, the activator can bind its binding site located upstream of the promoter (**a, b**). However, even in the absence of glucose, the operator turns on only in the presence of lactose when the lactose-bound repressor is in its inactive conformation (**b**). The operon is off when there is enough glucose in the bacterium since under these conditions, the activator is in its inactive conformation (no cyclic AMP) (**c, d**)

interacts with dozens of transcription regulators. This alone makes the remote location of the binding sites of many regulators unavoidable. Still, all of the regulators can interact with RNA polymerase simultaneously. Some of them do not contact the RNA polymerase but instead interact with the special large protein complex called a *mediator*, which binds to the RNA polymerase and various transcription

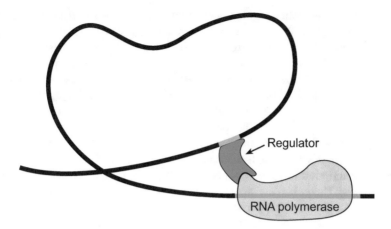

Fig. 5.3 Transcription regulation through DNA looping. The binding site for the regulator (green or gray in some formats) is separated from the target promoter along the double helix. Looping the DNA segment between the binding site of the regulator and promoter-bound RNA polymerase is necessary for the interaction between the regulator and the enzyme

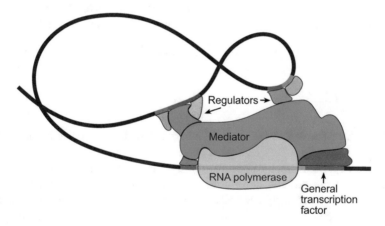

Fig. 5.4 Regulation of the transcription initiation in eukaryotes. The figure of an imaginary promoter shows the interaction of various regulators with the RNA polymerase and the mediator, a very large protein complex bound with the RNA polymerase. Each gene-specific regulator is bound with its specific binding site on DNA and with the transcription complex. A few dozen of proteins are involved in the regulation of an actual eukaryotic promoter

regulators (Fig. 5.4). The mediator serves as a bridge between transcription regulators and RNA polymerase. The eukaryotic transcription complex also includes a few regulators which have no sequence specificity. These regulators, called the *general transcription factors*, are required to begin transcription at any eukaryotic promoter.

Of course, transcription regulators can be useful only if the cell can control their activity. Eukaryotic cells use a few ways to implement such control. First, the presence of the regulator, like any protein, can be increased by its synthesis or reduced by proteolysis. This, of course, affects its activity. Second, the activity of the regulator can be affected by the binding of a ligand. Third, the activity can be changed by chemical modification of the regulator. Fourth, the activity of the regulator can be changed by binding to another protein. Fifth, the regulator entry to the nucleus can be temporarily blocked by another protein bound with the regulator.

In addition to the transcription regulators, eukaryotic transcription is regulated by the structure of chromatin (see Chap. 4). The complex of RNA polymerase and general transcription factors cannot be assembled on a promoter that is packed in nucleosomes. Therefore, the assembly requires that nucleosomes are made less stable or completely removed from the promoter. This gene-specific process is activated by the transcription activators, which can bind with a specific promoter and recruit nonspecific histone-modifying proteins to it. The proteins perform chemical modification of histones, causing changes in the chromatin structure and, correspondingly, altering the accessibility of the promoters. Similar to this, a promoter can be bound with transcription repressors, which can recruit other histone-modifying proteins. These proteins can change the segments of chromatin structure to make them less accessible to the transcription initiation or even make the genes completely silenced.

5.1.3   Cell Memory

Multicellular eukaryotes have very different types of specialized cells, and in each type of such cell, the sets of expressed genes are different. The majority of specialized cells experience many divisions during the organism's life (although some extremely specialized cells, like muscle cells and neurons, never divide). When they divide, they produce daughter cells of the same type. Thus, the daughter cells remember the patterns of gene expression responsible for their identity. How is the needed information passed through subsequent cell divisions?

The cell type depends, first of all, on the set of transcription regulators produced in that cell. Although the difference in expression involves several thousand genes, the cell type is specified by the expression of a combination of a few *master transcription regulators*. Each of these regulators interacts with many promoters and, therefore, strongly affects the set of expressed genes. A master transcription regulator can serve as an activator for some genes and as a repressor for others. To maintain the cell type, the set of the master transcription regulators expressed in the dividing cell has to be reproduced in the daughter cells. The way to achieve this is *positive feedback* in the regulation of a gene of a master transcription regulator. This means that a master transcription regulator has to serve as an activator for its gene. So, if the protein presents in the cell, its gene is expressed and new copies of the regulator are produced (Fig. 5.5). This means that the expressed gene of a master

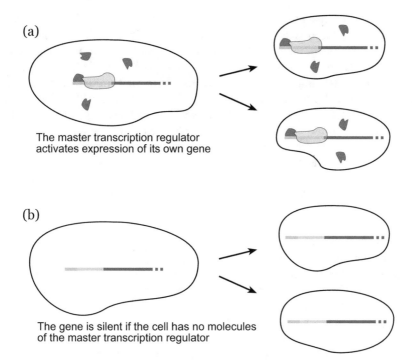

Fig. 5.5 Inheritance of the cell type by the positive feedback in gene regulation. In this example, a gene is activated by the protein coded by this gene. (**a**) Since the regulator molecules are distributed to both daughter cells, the gene will be expressed in both cells as well and the cells will produce more molecules of the regulator. (**b**) In the absence of the regulator in the cell, the gene is not expressed and it will not be expressed in the daughter cells. Such regulation of genes of the master transcription regulators provides the inheritance of the cell type

transcription regulator will be expressed in the daughter cells as well since the regulator molecules will go to each progeny cell during the cell division.

The positive feedback is the simplest regulation loop, which is transferred to progeny during the cell divisions. Another simple regulation loop provides *negative feedback* and allows the cell to keep the concentration of a regulator at a constant level. In this loop, the gene product serves as a repressor for this gene. Thus, if the product concentration grows above a certain level, the gene becomes more repressed and its activity reduces. This decreases the regulator concentration. If the regulator concentration drops below the normal level, the gene activity grows and stimulates the synthesis of the regulator molecules. All cells have various types of feedback loops, and many of them are combined in very complex networks. However, regardless of the network complexity, its state depends on the concentrations of the transcription regulators and various ligands which affect the activity of transcription regulators. During cell division, the concentrations of all these molecules are maintained, providing the mechanism of cell memory.

Another mechanism of cell memory is the methylation of cytosines in the genes which have to be repressed. The methylation occurs at 5′-CpG dinucleotide, which is paired with the same dinucleotide in the complementary strand. The methyl group added to a cytosine is exposed to the major groove of the double helix and, therefore, affects the binding of transcription regulators to the corresponding DNA segments. A pattern of DNA methylation at GpC steps can be inherited by the daughter cells and, therefore, suggests a simple mechanism of cell memory. If both cytosines are methylated in the CpG/CpG segment, then each of the new double-stranded DNA molecules receives one methylated cytosine during DNA replication (Fig. 5.6). After this, a special methylating enzyme acts on CpG steps, which are paired with CpG steps with already methylated cytosine. Thus, the methylation pattern of the parental DNA is reproduced in the daughter molecules. As a result, a gene repressed by the methylation in the parental DNA will be repressed in the progeny DNA as well.

Methylation of DNA creates obstacles for binding transcription activators and increases the affinity of the methylated promoters to transcription repressors. In addition to this, methylated DNA segments attract histone-modifying enzymes that convert chromatin into the strongly repressed state, enhancing the repressing effect of cytosine methylation.

The pattern of histone modification can be kept for many generations of particular cell types. To be transferred to the progeny cells, the histone modifications have to be reproduced during the cell division. Special, very large histone-modifying proteins are responsible for this task (see Chap. 4). Thus, the modification provides one more mechanism of cell memory.

An extreme example of gene silencing inherited during the division of somatic cells is the phenomenon called X chromosome inactivation. There are two X chromosomes in female mammals, while males have only one. Correspondingly, females have two copies of each X chromosome gene, while males have only one copy. To

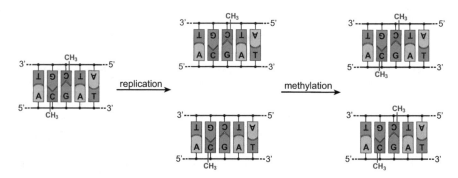

Fig. 5.6 Keeping the methylation pattern in the CpG steps during cell division. In the parental DNA sequence, CpG in one strand is paired with the same sequence in the other strand, and the cytosines in both strands are methylated. During DNA replication, each double helix receives one parental strand with methylated cytosine. Then, a special enzyme methylates the second cytosine in the half-methylated CpG steps

avoid the overexpression of these genes, mammals develop a mechanism of *dosage compensation*, which equalizes the number of active X chromosome genes between males and females. In female somatic cells, genes of one X chromosome, either the maternally inherited (X_m) or the paternally inherited (X_p), are completely inactivated by this mechanism. The initial choice of the X chromosome for the inactivation is random. However, once the choice is made, it maintains throughout all subsequent divisions of that cell and its progeny. Since the inactivation takes place in the embryo after several rounds of cell divisions, the growing organism receives a mosaic pattern of cells with active genes either from the X_m or X_p chromosome. The progeny of each cell tends to remain close together during later stages of development, so the body of an organism consists of cell clusters where the same X chromosome is active. An interesting illustration of this development is the black and orange coat coloration of some female cats (Fig. 5.7). It happens when one X chromosome carries a gene that produces orange hair color, while the same gene in the other X chromosome carries a black hair color. Male cats never have mosaic black and orange coloring since they carry only one X chromosome.

Modifications of chromatin, both DNA methylation and histone modifications, are reversible. The full proof of this was obtained in the experiments on replacing the nucleus of the egg cell with the nucleus of a specialized somatic cell. The egg with a replaced nucleus can be viable and capable of producing an adult organism containing all the necessary genetic information from just one parent. This procedure was first developed a few decades ago. In 1996, the first animal was grown after such a nucleus transfer, the sheep Dolly. This proves that all DNA and chromatin modifications in the specialized cells used in the experiments were reversed to form the normal state of chromatin in the obtained zygotes.

Fig. 5.7 Female cat with black and orange coat coloration. The coloration depends on a gene on the X chromosome. In female cells, only one of two X chromosomes is active, either maternally inherited or paternally inherited. The active X chromosomes are randomly clustered. Therefore, if two alleles of the gene code different colors, the female cat receives such mosaic coloration

5.2   Posttranscriptional Changes of mRNAs

5.2.1   Degradation of mRNA

Regulation of the transcription initiation described in the previous section only makes sense if the lifetime of mRNA molecules is relatively short. In another case, earlier produced mRNAs will be active, if not repressed, even if the expression of the corresponding genes is not needed for the cell. Therefore, all existing mRNA molecules are permanently digested. This digestion makes it possible to change the repertoire of active genes and proteins which they code, allowing the cells to react to changes in internal needs and external conditions. Usually, in bacteria, the average lifetime of mRNAs does not exceed 2 min, making them capable of quickly adjusting to changing conditions. The lifetime of mRNAs in eukaryotic cells is larger, around 30 min on average, although for some of them it can reach a few hours.

The degradation of mRNA is performed by special enzymes, nucleases. The activity of these nucleases is elaborately regulated. All eukaryotic mRNAs have a poly(A) tail at their 3′ end (≈200 nucleotides), which provides a necessary signal for the start of translation. mRNA digestion starts from a gradual shortening of the tail. The digestion of this tail serves as an internal clock for mRNA. When its length is reduced to about 25 nucleotides, other enzymes perform a faster degradation of the coding part of the mRNA molecule.

5.2.2   Alternative RNA Splicing

In many eukaryotic cells and, first of all, in mammalians, transcription of the genes produces only precursors of mRNAs. These precursors undergo RNA splicing, the procedure of removing large segments, introns, from the newly synthesized RNA molecules (see Chap. 4). The splicing can produce different mRNA molecules from the same gene by removing different introns from the pre-mRNA (Fig. 5.8).

In some cases, the number of possible alternative patterns of the splicing is very large, in extreme cases in tens of thousands. Although the proteins coded by mRNA

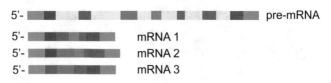

Fig. 5.8 Alternative splicing. A typical mammalian gene consists of exons, the segments of the protein-coding sequences, and introns, the segments which have to be cut out during the processing of pre-mRNA. Some exons can be removed from the pre-mRNA as well, and this results in the appearance of alternative mRNA molecules. Introns of the imaginary pre-mRNA are shown here in yellow, while exons are marked by different colors. Three mRNAs produced by alternative splicing of the pre-mRNA include different exons and, therefore, code different proteins

molecules produced by different splicing of the same pre-mRNA have similar functions, the difference between them can be very important. In particular, translation from alternative mRNAs can give different versions of a protein in different types of cells. The patterns of splicing are specified by specific loosely defined sequences at the ends of exons. In some instances, the splicing machine sliding along the pre-mRNA molecule can skip some of these sequences; in other cases, it ignores the others. The recognition of the splicing sequences by the splicing machine can be regulated by other proteins that bind with specific points of splicing.

5.2.3 Editing of mRNA

It seems that the repertoire of methods that regulate protein production is somehow excessive. In principle, cells could solve all problems of regulation by a smaller variety of methods, and some of these methods probably should be considered as patches to the major systems of gene expression regulation. Still, all existing systems of regulation are very important for a cell where they exist. An example of such an unusual system is the posttranscriptional editing of mRNA.

Editing is a process of enzymatic change of the RNA sequence. Two types of editing occur in mammalian mRNAs. The first one is the change of A to I (where I, inosine, can form the base pair with C when it interacts with the anticodon of tRNA). The second is the change of C to U (U forms the base pair with A). Both types of mRNA editing can result in the replacement of one amino acid by another in the corresponding protein. They can also create a new stop codon or affect the splicing of pre-mRNA. mRNAs from more than 1000 genes in humans are subjects of the editing. In mammalians, the sites of editing are specified by a special hairpin structure formed in the mRNA molecule (see Fig. 1.11). Other types of mRNA editing occur in some simpler organisms.

In many cases, the editing of mRNA introduces small but sometimes critically important changes in the coded protein. For example, a single amino acid replacement in the gene coding the protein of an ion channel makes the channel permeable for the ion. Without this change, mutant mice die in the early days of development. It remains unclear why the needed base could not be placed in the gene from the very beginning. Often, however, both unedited and edited versions of mRNA code proteins which properly perform their slightly different functions.

5.3 Suppression of Gene Expression by Small Noncoding RNA Molecules

It was discovered over the last 30 years that gene expression can also be regulated by small noncoding RNA molecules. These single-stranded RNAs (ssRNA) form so-called *RNA-induced silencing complexes* (RISC) with specific proteins. The complexes are capable of binding segments of target ssRNAs with complementarity to their RNA. Small ssRNA in the RISC complexes consists of 20–24 nucleotides and, in principle, can have any sequence. The binding of RISC with a target ssRNA inhibits its translation or even initiates the degradation of this ssRNA. The target ssRNA can be the cell's mRNA or the ssRNA of an invading virus. A remarkable feature of the RISC system is that the same set of proteins can form active complexes with guiding RNA of nearly any sequence. Therefore, they can be reprogrammed to suppress nearly any gene. In some organisms, like plants, RISC systems are very important as a self-defense mechanism against viral attacks. However, in mammalians, the system only regulates gene expression by suppressing the translation of the corresponding mRNAs.

There are three types of RISC systems that use different protein complexes and guiding RNA of different origins. These three types of systems are based on *small interfering RNA* (siRNA), *microRNA* (miRNA), and *PIWI-interacting RNA* (piRNA). Each of these systems includes different, although related, sets of proteins. Also, the guiding ssRNA molecules have different genesis in each of these systems (Fig. 5.9).

The system based on siRNA was discovered first, in the early 1990s. siRNA is double-stranded, 20–24 bp in length. It is produced by a special enzymatic complex from a longer dsRNA or small RNA hairpins. Any segment of double-stranded RNA of sufficient length can be converted into siRNA by cutting it to a specific length. In particular, it could be a viral dsRNA or hairpin structure with a stem of sufficient length formed by a single-stranded viral RNA, or replication intermediate of a viral genome. dsRNA of internal origin is not a common object in the cells, and special enzymes are needed for its production. It is obtained mainly by reannealing

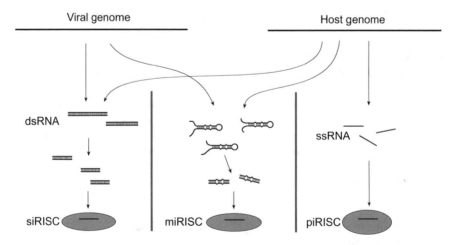

Fig. 5.9 The genesis of the small guiding ssRNA in the different RISC systems. siRNA is obtained from double-stranded RNA (dsRNA), either external or internal nature (left). miRNA is obtained from RNA hairpins, usually with mismatches in the stem (center). Although the hairpins are mainly transcripts from the host genome, they can have a viral origin as well. piRNA is obtained by transcription from the host genome in the germ cells; these RNA molecules do not pass the step of dsRNA

of mRNA molecules with complementary ssRNA transcribed on the antisense strands of certain genes. Further processing of those dsRNA results in the formation of siRNA molecules that are complementary to the mRNAs (or one of two strands of viral RNA). These siRNAs guide RISC binding to the target single-stranded RNA. siRNA-based RISC is much more active as an antiviral defense in plants than in animals. It was also found that siRNA plays an important role in the repression of transposons.

siRNA has become a very powerful research instrument. Indeed, the silencing of a particular gene can be achieved by injecting the corresponding synthetic dsRNA into the cells. This dsRNA is processed into siRNA by the RISC enzymatic machinery to guide the digestion or inhibition of the corresponding mRNA. siRNAs look very promising as therapeutic agents since they can silence practically any gene of internal or external origin with remarkable specificity. However, the delivery of synthetic dsRNA to the cells of living human organisms turned out to be a difficult problem. In general, cells do not allow foreign RNA to pass through their membrane. It took years to find ways to overcome numerous obstacles in the therapeutic use of siRNA. Only in 2018, the first siRNA-based drug was approved for clinical use.

Interestingly, many viruses developed an anti-siRNA system to protect their RNA from digestion by RISC. The key element of the system is a protein that binds RNA duplexes of a specific length, corresponding to the length of siRNA, regardless of their sequence.

The miRNA-based system has a lot of similarities with the system based on siRNA. miRNA is produced from RNA transcripts that fold back on themselves to form short hairpins. A specific feature of miRNA-based RISC is that full complementarity between miRNA and a target mRNA is not required for the miRNA-mediated transcription silencing, so a single miRNA can reduce the translation of hundreds of different genes. Around 1000 different miRNAs are coded in the human genome, and nearly all human viruses have sufficient complementarity with one or another miRNA. So, miRNA-based RISC can substantially reduce the translation of viral mRNAs, although they can hardly stop a viral invasion completely. The efficiency of the gene silencing is greater for mRNA, which has higher complementarity to miRNA. The enzymatic system that processes short RNA hairpins into miRNA can use viral RNA as well if it forms hairpins of sufficient size. Still, the main origin of miRNA is the host genome.

The main function of piRNAs is to silence transposons in the germ line of cells, and this role is maintained across animal species. piRNAs are processed from ssRNA coded by the cell genome. Although transposons make, probably, an important contribution to evolution, their movement must be minimized since very often it causes unwanted mutations in the host genome. The movement is only possible due to specific enzymes, transposases, coded by transposons. RISC systems based on piRNA repress the production of the transposases by silencing the corresponding mRNA molecules.

Mammals have hundreds of thousands of different piRNAs, many times more than their cells need to silence all transposons. The targets of piRNAs beyond silencing transposons remain unknown.

Chapter 6
DNA Replication, Its Fidelity, Mutations, and Repair

6.1 Replication of the Double Helix

It follows from the structure of the double helix that complementarity of bases in DNA strands provides remarkable regularity of the structure. Correspondingly, any deviation from the complementarity, like the formation of an AC pair rather than AT, disrupts this regularity of DNA structure. The disruptions increase the energy of the double helix. Therefore, the incorporation of only complementary AT and GC base pairs into the double helix is favorable energetically. This is why the requirement of complementarity suggests a very simple principle of doubling genetic information before cell division. However, a more detailed analysis shows that the process of DNA replication runs across many very difficult problems. These major problems of replication and the ways of solving them are considered below.

6.1.1 Directionality and Proofreading of DNA Synthesis

The first key problem which has to be solved during DNA replication is the necessity of extremely high accuracy of the process. Although the incorporation of non-complementary base pairs into the double helix increases its energy, the increase is rather moderate. The additional energy associated with a noncomplimentary base pair in the double helix is different for various pairs, but its average value is around 2 kcal/mol. This means that the spontaneous incorporation of an "incorrect" base into the double helix is about 30 times less probable than the incorporation of the correct one. Thus, the energetics of the double helix per se is capable of providing only very low accuracy of DNA duplication, about 1 error per 10 nucleotides in the new chain. Life requires incomparably higher precision of inheritance of the genetic information, and the cells manage to provide such remarkable precision.

© The Author(s), under exclusive license to Springer Nature Switzerland AG 2023
A. Vologodskii, *The Basics of Molecular Biology*,
https://doi.org/10.1007/978-3-031-19404-7_6

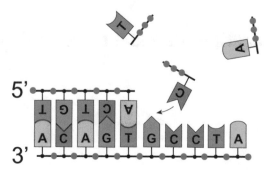

Fig. 6.1 Synthesis of the new strand during DNA replication. The parental strand (at the bottom) is used as a template. After a nucleotide triphosphate binds with the template at the 3′ end of the growing strand, the DNA polymerase tests whether the new base pair would fit the geometry of the double helix. If the test is successful, the enzyme attaches the nucleotide to the growing strand. The polymerization reaction also results in the release of diphosphate (not shown), which shifts the reaction equilibrium to polymerization. Phosphate groups are shown as small circles

In very general terms, the process of DNA replication consists of unwinding the complementary strands of the parental double helix and synthesizing, step by step, the new strands, using the old ones as templates (Fig. 6.1). The synthesis is catalyzed by the DNA polymerase, the major player in the complex of many proteins participating in the process. The DNA polymerase is capable of synthesizing the new strand only in the 5′-to-3′ direction. So, if a replication complex moves ahead from a particular point of the parental double helix, it cannot synthesize both new strands in the same way since the strands of the double helix have the antiparallel orientation. One possible solution is to have another DNA polymerase that performs synthesis in the 3′-to-5′ direction. Nature chose another way to handle this problem, however, because such an imaginary enzyme would reduce the precision of the synthesis. To understand this, we have to consider the reaction in more detail.

At each step of the synthesis, attaching a new incoming nucleotide to the growing chain is accompanied by the release of diphosphate. The released diphosphate makes the reaction highly favorable energetically. After each step of the synthesis, the DNA polymerase translocates the growing double helix and tests one more time that the correct nucleotide was incorporated, and the new double helix maintains its regular structure. If this test fails, the enzyme detaches the mismatched nucleotide and pulls back the growing double helix to attach the correct nucleotide. The back step produces a nucleotide monophosphate rather than a nucleotide triphosphate (Fig. 6.2a). This is not a problem for the synthesis in the 5′-to-3′ direction since a new correct nucleotide carries the triphosphate needed for the polymerization (see Fig. 6.1). However, the back step would abort the polymerization which proceeds in the 3′-to-5′ direction. Indeed, such synthesis requires the presence of triphosphate at the growing end of the new strand, as the energy source for the next step of polymerization. The back step would only leave the monophosphate at the chain end terminating the synthesis (Fig. 6.2b). So, the 3′-to-5′ polymerization is incompatible with the back steps, and this type of sequence correction in the new strand would be impossible.

The backtracking proofreading described above is not the first test of synthesis accuracy. The first test happens before the DNA polymerase forms the chemical bond

(a)

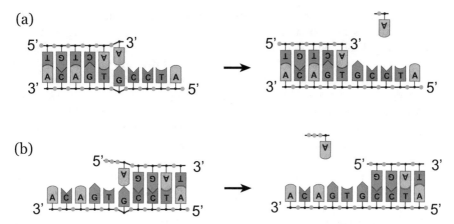

(b)

Fig. 6.2 The back step during the synthesis of the new strand. (**a**) If the synthesis goes in the 5′-to-3′ direction (as it does), the removal of incorrect nucleotide returns the end of the growing strand into its initial state. There is no obstacle to attaching the next correct nucleotide after the back step. The detached nucleotide monophosphate will be recycled. (**b**) If the synthesis goes in the 3′-to-5′ direction (which does not happen), the removal of the incorrect nucleotide does not return the end of the growing strand into its initial state, with triphosphate at the end. The synthesis would be aborted in this case

between the new incoming nucleotide and the growing chain. This preliminary test is also based on the general geometrical requirement of the base pair formed by the bound nucleotide. The failure of the test blocks the nucleotide attachment, it dissociates from the template site, and a new nucleotide binds the site. This first test is quite efficient; it allows, on average, one error per 10^5 added nucleotides. The backtracking proofreading reduces the probability of a mistake to 1 per 10^7 added nucleotides. To reach an even higher accuracy of replication, 1 mistake per 10^{10} steps, the cell uses post-replicational correction. The system of special enzymes detects mismatches by sensing the misfit in the double helix and excises the DNA segment containing the mismatch in the newly synthesized strand. Then the excised portion of the strand is resynthesized. Of course, during this correction step, the system has to distinguish between old and new strands to make the excision in the new one. This is achieved by making temporary chemical marks in the new or old DNA strands.

These three proofreading systems are essentially the same throughout all domains of life, which emphasizes their critical importance. In mammals, these systems provide an accuracy of 1 mistake per 10^{10} added nucleotides. We will see below, however, that maintaining the correct DNA sequences requires permanent repair not just during DNA replication but at all moments of the cell's life.

6.1.2   Replication of Leading and Lagging Strands

DNA replication starts when the replication machinery is assembled at a special region of the genome, called *replication origin*. Bacteria have only one replication origin, which extends over a few hundred base pairs. Each chromosome of eukaryotes has many origins of replication (about 1000 per chromosome), and each of them consists of thousands of base pairs. The sequence of the origins is not strictly defined, although, in bacteria, it has a higher fraction of AT base pairs. AT base pairs are less stable than GC base pairs, so it is easier to unwind the double helix in AT-rich regions. Local DNA unwinding, assisted by a few proteins, is a necessary step in the assembly of replication complexes at the origin. Since the DNA polymerase can synthesize the new single strands only in the 5'-to-3' direction, there is no simple way to perform the synthesis for both strands simultaneously. To overcome this problem, the DNA polymerase synthesizes continuously only one strand, for which the movement of the replication complex corresponds to the 5'-to-3' direction in the growing strand. The other new strand is synthesized as fragments (Fig. 6.3). The length of the fragments (called *Okazaki fragments*) is about 200 nucleotides in eukaryotic cells and about 2000 nucleotides in bacteria.

The elaborated proofreading mechanism in the DNA polymerase is incompatible with the start of a new strand, and the enzyme can only elongate it. Therefore, another special enzyme, *DNA primase*, produces short primers (oligonucleotides) which are base-paired with the parental strands. The primers allow the DNA polymerase to start the synthesis by elongating them (see Fig. 6.3). Of course, only one primer per replication origin is needed on the leading strand, while the synthesis of the lagging strand requires a primer for each Okazaki fragment. Since a polymerase that is capable of starting the synthesis cannot be accurate enough, the primers have

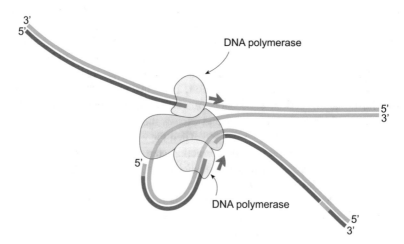

Fig. 6.3 Synthesis of the new strands during DNA replication. The parental DNA (orange) is unwound, and both of its strands are used as templates to synthesize the new strands (brown). Since the DNA polymerase can perform synthesis only in the 5'-to-3' direction, the synthesis proceeds continuously on the leading strand, while on the lagging strand it produces fragments (its direction is shown by green arrows). RNA primers (green) are needed for the DNA polymerase to start DNA synthesis. Later these primers are replaced by DNA segments and the fragments are joined in a continuous chain by the DNA ligase. The replication complex consists of many enzymes, but for simplicity, all of them, except the DNA polymerases, are shown as a large blue object. Only one of two replication complexes, diverging from the replication origin, is shown

to be later removed and replaced by DNA segments synthesized by the DNA polymerase. To mark the primers for removal, the *DNA primase* makes them from ribonucleotides. When the synthesis of the lagging strand is complete, the fragments are joined together by the special enzyme, DNA ligase.

6.1.3 Telomeres and the Organism Aging

The necessity of the primers creates another problem in DNA replication, however. The RNA primer located at the 5′ end of the new strand cannot be replaced by the DNA segment because there is no 3′ end available for the DNA polymerase. The cell needs a mechanism to overcome this problem, to avoid losing the ends of the chromosomes each time it divides.

Bacteria solve this problem by closing their DNA molecules into a circle, so they do not have ends. Eukaryotes use instead special segments at DNA ends, which form *telomeres*. The telomeres contain multiple repeats of a simple short sequence, like GGGTTA in humans, which is repeated hundreds of times at each telomere. During DNA replication, telomeres at the 3′ ends of the parental DNA are first extended by a special enzyme, *telomerase*. The telomerase uses its internal template which corresponds to a single repeat of the telomere sequence. The extended ends of parental DNA are replicated in the regular way, although it results in their shortening in the new strands. Thus, the telomere sequence grows and shrinks during each replication cycle. These two processes are only approximately balanced, so the chromosome ends contain a variable number of telomeric repeats. The balance depends on telomerase activity. In the majority of somatic cells, telomeres shrink, on average, at each cell division. There are solid reasons to believe that such shrinking of telomeres restricts the number of divisions of such somatic cells and, probably, restricts the life span of the organism. The restriction makes sense since each division requires DNA replication, which is the main source of mutations in daughter DNA molecules. This accumulation of mutations becomes, eventually, incompatible with the normal life of somatic cells. The shrinking does not occur in germ cells, where DNA replication is much more accurate (see below).

6.1.4 Replication Complex and Epigenetic Inheritance

The replication complex represents a large machine that consists of about a dozen different proteins. There are proteins that unwind the double helix, consuming the energy of ATP hydrolysis and keeping it in the unwound state. The strands of the double helix are tightly interwound and the same interwinding would remain between the newly made double helices since the parental strands remain intact during the replication. Keeping the same interwinding is impossible, however, due to the very high bending rigidity of double-stranded DNA. In linear chromosomes, rotation of DNA ends could solve the problem, but such rotation in viscous media is extremely slow. Therefore, the replication stops at the very beginning if it is not assisted by special enzymes, DNA topoisomerases (see Sect. 2.3). Topoisomerases catalyze the passage of one DNA segment through another. This reaction resolves all topological problems which appear in DNA functioning. Therefore, the replication complex also includes DNA topoisomerases.

The replication complex in eukaryotic cells has one more important set of proteins involved in the initial chromosome modification, as mentioned in Chap. 4. The goal of these proteins is to reproduce the parental chromosome structure in the newly made chromosomes. For this purpose, the replication is tightly coupled with the formation of nucleosomes and their remodeling. Old histone proteins are released from DNA and the nucleosome core is disassembled before the next Okazaki fragment is synthesized. The released proteins are used in the nucleosome duplication which follows the synthesis of the Okazaki fragment. Some of these proteins can be chemically modified. In such cases, special enzymes of the replication complex reproduce the modification of the newly recruited nucleosome proteins. Eventually, the chromatin remodeling complexes duplicate the local structures of the parental chromosomes. In this way, chromosome duplication preserves the type of the dividing cell. Thus, the coupling between DNA replication and chromatin modification provides a mechanism for epigenetic inheritance.

6.2   DNA Damage and Repair

DNA is a chemically stable polymer. Bases, the sequence of which carries genetic information, are hidden inside the double helix and therefore protected from chemical reagents that exist in the cell environment. Still, it suffers constant chemical damage from various active substances in the environment, radiation, and heat. In a human cell, tens of thousands of chemical changes in DNA molecules appear every day. However, nearly all of them are successfully repaired, so the permanent changes, called mutations, accumulate very slowly. Such a very high rate of reparation is due to the structure of the double helix where information recorded in one strand is duplicated in the complementary strand. Many hundreds of proteins are involved in DNA repair in a human cell. Some of these proteins can be found in all species, from bacteria to vertebrates. Still, sometimes the repair system misses a structural change, and DNA replication results in the appearance of a mutation (Fig. 6.4).

The great majority of chemical modifications can be easily identified since they do not result in the formation of a canonical nucleotide from another one. Also, a very small fraction of them involves both bases of a pair. Therefore, the repair system can use the intact strand as a guide for the repair. The repair starts with removing the damaged nucleotide from the chain. For example, in the case of depurination, the most frequent chemical damage of DNA, the base pair loses a purine attached to the deoxyribose. A special endonuclease recognizes the place of the base loss and cuts sugar and

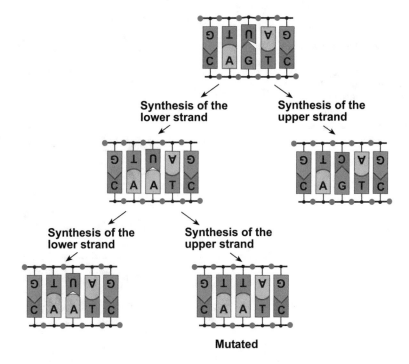

Mutated

Fig. 6.4 The appearance of a point mutation in DNA. A single chemical reaction can convert cytosine into uracil, causing the appearance of the GU base pair (upper row). If the change is not corrected by the repair system, the first round of replication results in one DNA molecule with an AU base pair, which is well accommodated in the double helix, and one DNA with the original GC base pair. The second round of replication results in the appearance of one mutated DNA where the original GC base pair is replaced by the AT base pair

phosphate from the damaged strand (Fig. 6.5). Then the DNA polymerase fills the gap, and DNA ligase reseals the nick. A similar pathway of events is used to repair other single-nucleotide damages. Some damages involve more than one nucleotide, like pyrimidine dimers caused by sun radiation. In these dimers, two adjacent pyrimidines, located in the same strand, are bound together by chemical bonds. To repair such extended damages, a set of enzymes cut and remove the single-stranded segment of a few nucleotides in length, resynthesize it, and ligate the nick.

The more complex situation appears when the damage involves methylated cytosine. In vertebrates, such methylated cytosines often appear in silenced genes, in CpG steps. The problem here is that the accidental deamination of the methylated cytosine produces the natural nucleotide thymine, resulting in the TG base pair. So, the repair system has to decide which strand contains an incorrect base. To address this problem, the cells have a special enzyme, DNA glycosylase, that recognizes TG base pairs involving T in the sequence TpG and removes the T (Fig. 6.6). This is not a very reliable strategy, however, and methylated cytosines in CpG steps remain to be the main mutation spots in mammalian genomes.

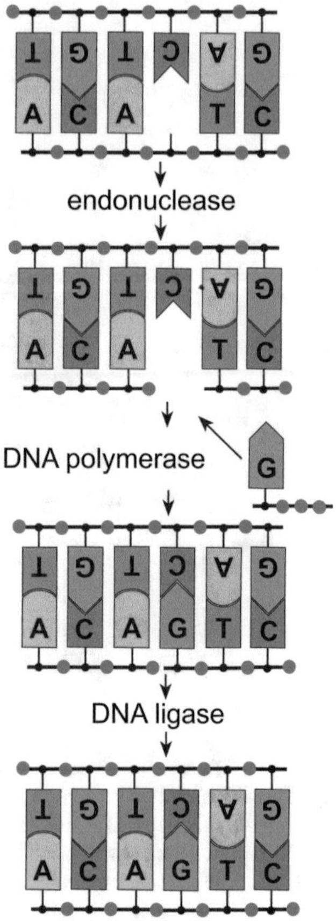

Fig. 6.5 Repairing of depurinated base pair. A special endonuclease finds the damage and makes a single-nucleotide gap in the damaged strand. Then the DNA polymerase adds a new nucleotide at the place and the DNA ligase seals the nick

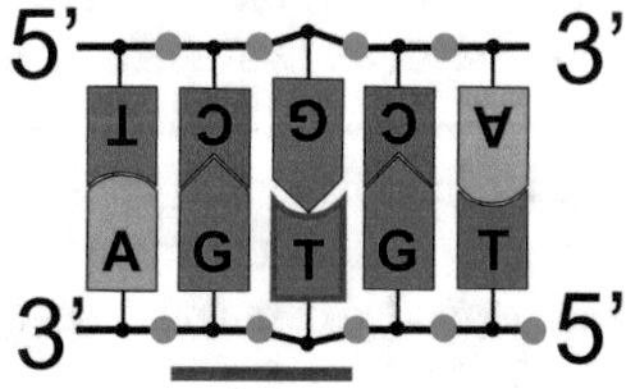

Fig. 6.6 Thymine formed from methylated cytosine in the CpG step. The repair system looks for the TG base pair in the TpG step (marked by underline) and replaces T with C

The most dangerous chemical damage to DNA is the double-stranded break. In this case, there is no intact template for the repair. The breaks are caused by ionizing radiation, various chemical reagents, and replication errors. The main system which repairs double-stranded breaks simply joins the newly formed ends of the double helix. Since the ends are usually damaged, rejoining them requires removing a few nucleotides at the ends. Therefore, this way of repair results in a loss of 1–3 base pairs at the repair site. Of course, this is a very serious and potentially damaging mutation. It seems that such lesions do not happen too often, although the issue requires further investigation.

There is another, more complex but accurate way to repair double-stranded breaks. During a relatively short period of cell life, each chromosome is represented by two identical sister chromatids. This is a period after DNA replication but before cell division. If a double-stranded break happens in this period, the intact sister chromatid can be used as a template to repair the broken one. The corresponding chain of reactions represents an example of *homologous recombination*. Homologous recombination is not just a repair mechanism for double-stranded breaks. It also allows the exchange of double-stranded DNA segments between the pairs of homologous chromosomes. This recombination is a very complex process involving many enzymes. We will not discuss its details here.

6.3 Evolution of Species

We do not know how life started on Earth. Indeed, even the simplest prokaryotic cell has so complex an organization that it is nearly impossible to imagine how it could appear as a result of random chemical reactions. Probably, there were simpler forms of life that served as intermediates, but we know nothing about this. It seems that no traces were left of such forms, so the origin of life will remain a mystery for a visible future. Our understanding of the subsequent development of life, called evolution, is much better.

6.3.1 The Theory of Evolution

The fundamental fact is that the cells of all organisms, despite their incredible variety, have strikingly identical organization. In all of them, genetic information is coded in DNA. It is transmitted to RNA molecules which are then used to synthesize proteins. The sequences of amino acids in proteins are coded by the same code in all organisms. The synthesis of proteins in all organisms is performed in the ribosomes with the tRNA adaptors (see Sect. 1.7). This does not mean that only this form of life existed, but even if there were other forms, they were extinct without traces.

This practically identical organization of cells suggests that all current and extinct species are dissent of one, which existed a long time ago. Charles Darwin was the first who formulated this idea in his work "The Origin of Species" published first in 1859. The book is now widely considered the most influential scientific book of all time. Remarkably, it was published nearly a century before the era of molecular biology, which made Darwin's theory undeniable. The theory stated that all species on Earth descended from a single common origin through evolution and branching. Darwin concluded that all species consist of slightly different individuals, and these individuals can slowly change over time. Some individuals can become better adjusted to current conditions, so they receive advantages over others in competition for natural resources and have better chances for survival. The difference between individuals can be transferred to the descendants. Eventually, over many generations, the process can result in the appearance of a new group of individuals inside the current one. If the new group has advantages over the original one, it can replace it. Alternatively, the original group can be split into two with different features.

Today, we know that inheritable changes in individuals, a key concept of Darwin's theory, are caused by mutations in their DNA. Mutations are a driving force of evolution. Some fraction of them is beneficial for the organism and, as such, gives a reproductive advantage to individuals who carry such mutations. However, the fraction of beneficial mutations is extremely small. A much larger fraction of mutations are deleterious. The majority of mutations in mammals are neutral, first of all, because only a small part of the genome constitutes the coding segments of DNA. Over evolution, individuals with deleterious mutations that perturb important functions are eliminated by purifying selection. This process maintains the fitness of the population.

6.3.2   The Evolution Tree

Remarkable progress in DNA sequencing made it possible to obtain the genomes of many organisms. Comparing their sequences has brought a lot of very important information about life and its evolution on our planet. Today, we know that the nucleotide sequences of many genes were highly conserved through evolution. Unmistakable similarity, or homology, is seen between many genes of humans and numerous other organisms. This homology extends not only to the genes of remote vertebrates but to the genes of flies, yeast, and even bacteria and archaea. The current molecular data strongly support Darwin's suggestion that all living organisms originated from a single remote ancestor (or a colony of ancestors). These data extended Darwin's evolution theory by including in it two major domains of life, bacteria and archaea, while Darwin considered only animals.

Mutations in DNA are accumulating with time, and therefore, they can be used as a biological clock. There is no doubt that apes and humans have a common ancestor, in particular, because their genomes are so similar. The detailed comparison shows that the genomes of chimpanzees and humans are closer one to another than those of humans and orangutans. This means that humans and chimpanzees diverged from a common ancestor later in their evolution than humans and orangutans from their common ancestor. If we know the speed of mutation accumulations in a particular gene, we can estimate, by comparing DNA sequences of the gene in two species, when they diverged in the evolution. Some proteins interact with many other proteins and nucleic acids. With high probability, changes in their structure would compromise one or another of these interactions; therefore, their structures are highly conserved. Correspondingly, the genes of these proteins evolve very slowly through evolution, while other proteins accumulate mutations faster. Due to this difference in the speed of evolution of various genes, it is usually possible to find a DNA segment with a sufficient amount of differences for any pair of species. This allows for determining the time of their divergence (Fig. 6.7). Many very interesting events in evolution can be tracked by comparison of DNA sequences of various genomes. The analysis of genomes transformed evolutionary biology.

The divergence of the major domains of life, bacteria, archaea, and eukaryotes, was deduced from the sequences of ribosomal RNA and ribosomal proteins. It was found that bacteria and archaea diverged about 3.2–3.5 billion years ago, while eukaryotes diverged from archaea about 2 billion years ago.

6.3.3   Conserved Regions of Eukaryotic Genomes

Very important conclusions were derived from the comparison of human genomes with the corresponding genomes of other mammals. In general, genomes of evolutionary close species exhibit large similarities. The degree of this similarity is very different, however, for exons and introns. This is what one should expect since mutations in introns are mainly unimportant for an organism, while changes in exons are usually deleterious and as such are a subject of the purifying selection. The exons of genes represent an example of conserved sequences that change very slowly through evolution. A natural generalization of this observation is the conclusion that all DNA segments which are important for the life of an organism should

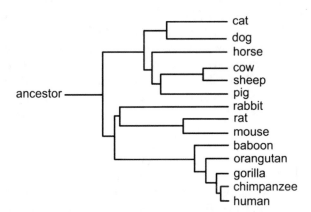

Fig. 6.7 The evolutionary tree for some mammals deduced from genome comparison. The length of the horizontal lines is proportional to the time since the lineages of the two species split. This tree of life can be extended to include all current species, including bacteria and archaea. All mammals diverged from a common ancestor about 400 million years ago

be in conserved regions of the genome. This conclusion became one of the major principles of genomics, a field of biology focusing on the structure of genomes.

The comparative analysis shows that about 5% of the human genome belongs to the conserved regions. These regions include protein-coding segments of genes, genes that code RNA molecules with known functions, DNA segments that are involved in transcription regulation, and DNA replication. However, in the human genome, the above DNA segments constitute less than half of the conserved regions (less than 2.5%), while the functions of the other part remain unknown. This striking finding shows that many things about our life remain to be discovered.

6.3.4   Genomes and the Human History

Over the past two decades, researchers have learned to reconstruct DNA sequences from the remains of individuals who lived thousands and sometimes hundreds of thousands of years ago. This has not been an easy task, because ancient DNA is highly degraded, and only short fragments of long molecules can be extracted from the remains. Another difficulty is the excessive contamination of the excavated samples by bacterial DNA. In the end, all these difficulties were overcome. The approach has been successfully applied, first of all, to remains of human DNA dated thousands and tens of thousands of years ago. The obtained findings have changed our understanding of human history. Although there is little hope of sequencing the DNA of organisms that died millions of years ago, in some cases indirect reconstruction of their genes becomes possible. The approach to this reconstruction is based on comparing the genes of several modern species. It allows for obtaining, indirectly, the sequence of their common ancestor. In this way, we can reconstruct genomes of organisms which became extinct tens of millions of years ago. The sequencing of ancient DNA has brought a lot of new information about the history of humanity, above all about human migration. It has expanded our knowledge of the past by many thousands of years when other data were very limited. In particular, the analysis confirmed and enriched some of the fossil data on the early population of Europe. It was found that about 600,000 years ago, Neanderthals and our ancestors, called modern humans, diverged into separate lineages. About 200,000 years ago, Neanderthals settled in Europe by migrating from Africa. Modern humans began moving into Europe about 50,000 years ago. For 10,000–20,000 years, they lived side by side with Neanderthals. A comparison of the genomes of the two species unexpectedly showed that there was interbreeding between Neanderthals and modern humans at this time. This means that the divergence between the two lineages for almost 600,000 years was not sufficient to completely block their admixing. This was the first example showing that branches of the evolutionary tree can not only diverge but can also rejoin. However, this admixing was not without hardship. Eventually, about 30,000 years ago, Neanderthals became extinct, for reasons which we still do not know. Now our genomes contain only 2–4% of Neanderthal's genes.

About 4500 years ago, another dramatic population change happened in Europe. People from the Eurasian steppe north of the Black Sea, called Yamnaya people (after their characteristic burial tradition), invaded all of Central, Northern, and Eastern Europe and the British Islands. Their genomes were notably different from the genomes of people who had occupied these lands, allowing reliable tracing of the population changes. Numerous data show that Yamnaya people nearly completely replaced the population of these parts of Europe. Their fast invasion into Europe was possible due to important technological advantages of their culture, first of all, the wheels which were used in carts and wagons. These were probably the first wheels in the world. The Yamnaya people were herders and the wheels were a key factor for necessary movement over extended territories. Although in later times other invaders contributed to the genetic landscape of Europe, the Yamnaya people made the major input into modern genotypes of Northern and Eastern Europe. The high mobility of the Yamnaya people also allowed them to spread over Central and South Asia. Some data suggest that the culture of the Yamnaya people may be the base of Indo-European languages, although genome analysis alone is hardly sufficient to prove it.

6.4 Accumulation of Mutations in Human Individuals and Populations

The quantitative picture of mutation accumulation is shaping only now, due to the revolution in DNA sequencing. In particular, the possibility to sequence DNA from a single cell, developed recently, became especially important in the study of mutation accumulation in eukaryotic cells. Of course, the major attention of researchers has been dedicated to mutations in humans, and this is the central topic of the paragraph.

Considering the accumulation of mutations in mammals, we should distinguish between somatic and germ cells. Mutations in the somatic cells cannot be inherited by the offspring. They are only transmitted to the next generation of the same line of somatic cells of the same organism. It does not mean that mutations in somatic cells are harmless. They could cause the formation of cancer cells or cells with compromised functions. Although the majority of somatic cells with seriously harmful mutations die, a small fraction of them causes serious problems for the organism. On the other hand, any mutation which happened in a germ cell is transmitted to the gametes and, correspondingly, can appear in the offspring.

Discussing the mutation rate, we have to normalize it for a time interval, or per some critical event. Chemical modifications of DNA bases do not cause, as a rule, substitutions of one canonical base pair by another canonical one. The mutated DNA molecules appear only after one or two rounds of replication when a modified base causes a replication error (see Fig. 6.4). However, the rate of mutations is often normalized by the number of cell divisions. Another way, especially appropriate for the germ cells, is to specify the number of new mutations per generation of the species.

6.4.1 Mutations in Somatic Cells

Only in recent years, the rate of mutations in human somatic cells was determined accurately. It was found that the rate is close to $3 \cdot 10^{-9}$ mutations per bp per cell division. Nearly all of these mutations are replacements of single base pairs by different ones, like C·G base pair by T·A, or C·G by G·C. Since the human genome consists of $3 \cdot 10^9$ bp, this rate means that about ten new mutations are accumulated in the genome over each cell cycle. Since only about 5% of the human genome contains critical information for cell development, mutations accumulated over a single cycle of a somatic cell will be tolerable in most cases. However, over the human life span, the division of many somatic cells occurs more than a 100 times, resulting in a 1000 mutations. It is not so clear how cells manage to survive with such an amount of mutations.

It seems that the accumulation of mutations in somatic cells is the main reason for organism aging. It was found that the rate of accumulating somatic mutations varies widely for different mammals. However, at the end of the life span, all investigated species have approximately the same number of new mutations in the DNA of their somatic cells. We can conclude that a notably larger number of new mutations are incompatible with the life of an individual.

6.4.2 Mutations in Germ Cells and the Age of Parents

Mutations in germ cells are transmitted to offspring and therefore are especially important. The recent data show that the rate of new mutations is much lower in the human germ cells than in the somatic cells. It turned out that the rate equals $3 \cdot 10^{-11}$ per bp per cell division, a 100 times lower than the mutation rate per cell division in human somatic cells. It means, probably, that the genome is repaired more carefully in the germ cells. In the case of germ cells, it is more interesting to consider the number of new mutations per generation. It turned out that this number depends on the parents' age. The number of mutations transmitted to a child increases linearly both with the age of the father and with the age of the mother. However, at all ages, the father brings three to four times more mutations than the mother (Fig. 6.8). It is understandable since the father's germ cells undergo many divisions over his life span, while the mother's ones do not. As it was noted above, DNA replication during cell division is the major source of mutations. Overall, a child obtains 50–100 new mutations from his/her parents. Correspondingly, 2–5 new mutations fall into 5% of important, conserved DNA segments. The majority of the latter mutations are deleterious, at least slightly.

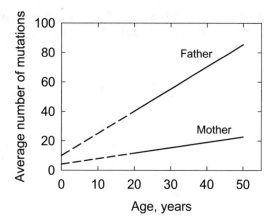

Fig. 6.8 The average number of new mutations in human germ cells which are transmitted to a child

Such a relatively high rate of accumulating mutations in new generations worsens the overall genetic fitness of the human population. Of cause, such a rate is not a specific feature of humans. It seems even higher in mice, for example. However, in wild nature, the fitness of a species is maintained by purifying selection, a factor that is weakened in the modern human population due to a high level of food supply, greatly improved living conditions, and health care. Therefore, the genetic fitness of the human population is worsening over the last couple of centuries. The increasing rate of mutation-related diseases, like cancer or neurological diseases, is related to the accumulation of deleterious mutations.

The data presented in Fig. 6.7 show that healthy younger parents and especially younger fathers have a better chance to have healthier children. It does not mean, however, that the genetic fitness of the population in a relatively isolated reproductive group is worsening faster if the age of parents in this group is higher, as some researchers concluded. The data in Fig. 6.7 show the opposite. One can see from the figure that there is an age-independent constant contribution to the average number of new mutations in the next generation (ten mutations for men and four for women). Therefore, the total average number of new mutations accumulated in the group over an extended period will be larger for a larger number of generations that appeared in this period. Correspondingly, the genetic fitness of the group will decline faster if more generations appear over the period. Thus, the higher age of parents is beneficial for the group's genetic fitness. The increasing frequency of some mutation-related diseases in isolated populations is due to a combination of the common decline of genetic fitness and a negative effect of small isolated reproductive groups. The latter factor increases the probability that a child will have the same mutation in both alleles of a gene, so all products of the gene will be affected.

Although our genetic fitness is declining, the future of humankind does not look so bad. Great advances in gene therapy allow us to hope that in the visible future, it will be possible to correct many mutated genes in somatic cells, to improve the fitness of individuals with harmful mutations. It is more difficult to correct genes in zygotes, which would result in healthier progeny (see Chap. 4). This task requires more technological advances because errors in gene therapy must be excluded in this case. However, the problem does not seem unsolvable.

6.4.3 Tracing the Past and Predicting the Future

The rate of accumulation of new mutations in the human genome, 50–100 mutations per generation, is relatively low if we consider the genome size. Therefore, the genomes of close relatives differ only slightly. On the other hand, up to 0.5% or more than 10^7 of all bases in the genome can be different for two persons who originated from common ancestors a very long time ago. Thus, the mutations in the genome carry information about many generations of ancestors of each individual. Correspondingly, the comparison of genomes allows us to trace the origin of individuals.

Comparative analysis of genomes became a booming direction in modern biology. From this analysis, we can learn a lot about our individual history, and many commercial companies offer this kind of genetic analysis. Over the last few years, researchers learn how to sequence DNA extracted from bones of ancient people who lived tens of thousands of years ago, and comparative analysis of the sequenced genomes revolutionized archeology. These studies allowed researchers to trace human migrations over thousands and tens of thousands of years. No other method could give this kind of information with comparable precision. This very exciting topic is behind the scope of the book, however.

We are all different, and the difference in our genomes makes a major contribution to variations in our phenotypes. There is no need to argue that the colors of eyes and hair are specified in the genes. It became proven a long time ago, however, that many other features of individuals strongly depend on the genome as well. It is true about the physical abilities of individuals, their health, intellectual abilities, and features of their characters as well. It does not mean that everything is completely specified in the genome, but the influence of our DNA is very important. It was not possible before recently to say what alternations in the genome are responsible for a particular feature, except for some genetic illnesses. Although we are still far from a complete understanding of the genome texts, we can now use an empirical approach to establish correlations between the genome alterations and specific features of individuals. For example, we can choose two large groups of individuals, very tall ones and very short persons, and compare their genomes. Of course, there are many differences in their DNA that are related to varied features. However, we can search for common differences between the two groups, and if the groups are large enough, these common differences have to be related to height. This kind of study of genetic determinants of various features is in progress now and they are bringing many extremely important results. The results are refined as more data for the statistical analysis becomes available. These will certainly bring very serious consequences to the future of humankind.

Chapter 7
Transduction of Signals

7.1 The General Principle

All cells change their behavior during their lifetime. These changes are the responses to changing external conditions or can be caused by the cell's development. Correspondingly, all responses are initiated by specific signals that the cell receives from other cells or generates itself inside its own body. In this chapter, we consider the *signal transductions*, signaling processes that start from the cell exterior. Internal signals have, in general, similar mechanisms of transmission, although in this case, there is no need for a signal to pass through the cell membrane. The major features of the signaling pathways are conserved through all domains of life.

The communication between the cells of multicellular organisms forms an extremely complex elaborated system. Still, we can outline some general principles of the process.

Communication between the cells is mediated by extracellular signal molecules that do not penetrate, usually inside the cells which receive the signals. Instead, the signal molecules are bound by specific *receptors* located at the surface of the receiving cell. The receptors are transmembrane proteins that have highly specific binding sites for the signal molecules (Fig. 7.1). The binding sites are exposed to the extracellular space, while the other end of receptors is located in the cytoplasm. The binding of the signal molecule by the receptor causes an allosteric change in its conformation. This change greatly enhances the affinity of the receptor to another signaling protein located in the cytoplasm. Binding this protein with the receptor causes a chemical modification of the protein and its conformational change. The modified first signaling protein binds to the second signaling molecule, usually a protein, and induces its conformational transformation. After a few such steps, the signal reaches the target protein. This target protein can be a transcription regulator or a metabolic enzyme. The received signal results in chemical modification and a change of activity of the target protein. Thus, the signal is transmitted to the target protein through a chain of signaling proteins.

A. Vologodskii, *The Basics of Molecular Biology*,
https://doi.org/10.1007/978-3-031-19404-7_7

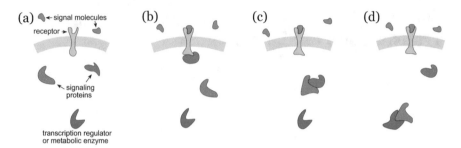

Fig. 7.1 Simple signaling pathway. (**a**) The chain of the signal transduction events starts when the signal molecules appear in the vicinity of the cell with the corresponding receptor. (**b**) A signal molecule binds with the receptor and changes its conformation. This conformational change makes the interior end of the receptor capable of binding to the first signaling protein. The binding causes a chemical modification of the protein and is followed by a change in the protein conformation. (**c**) The first signaling protein dissociates from the receptor and binds to the second signaling protein. The binding induces a chemical modification and a conformational change in the second signaling protein. (**d**) The second signaling protein binds the target protein causing its modification and a conformational change

The cells usually use chains of signaling proteins rather than just one intracellular protein for the transduction of each signal. The reason for this is that there is no one-to-one correspondence between receptors and target proteins. At each step of signal transduction, more than one signaling protein may serve as a signal sender, and more than one may receive the signal. The signaling chains form an extremely complex network that can integrate signals from many receptors and deliver them to different targets. So, each link of the signaling pathway can be considered as a branch point of the signaling network. These signaling networks are different in different types of cells of the same organism.

Receptors that bind signal molecules have remarkable specificity. Each cell has many different receptors embedded into its membrane, and the sets of receptors that are embedded in the membrane of different types of cells of a multicellular organism are different. More than 1500 receptors are coded in the human genome.

The type of signaling pathway shown in Fig. 7.1 is not the only one that exists in multicellular organisms. Ion channels can also represent receptors that change their conformation by binding signal molecules. In this case, the signaling pathway is short—the channels bind the signal molecule and open or close themselves for the flow of specific ions. Also, some receptors are sensitive to changes in external pressure or react to light, like the receptors in our eyes.

7.2 How Receptors Receive Signal Molecules

Nearly every cell is capable of secreting signal molecules. In this way, cells communicate with each other in multicellular organisms. Even single-cell organisms release signals into the surrounding space for the neighboring cells. There are a few ways of delivering signals to the receptors of the target cells.

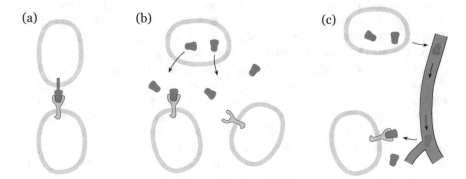

Fig. 7.2 Various ways of obtaining signals by the receiving cells. (**a**) The signal molecules remain bound to the membrane of the cell generating the signal. Direct contact with the receiving cell is required for signal transduction. (**b**) The signal molecules are secreted into the intercellular space by the cell, which sends the signal. The molecules diffuse to the receptors of receiving cells. (**c**) The signal molecules, *hormones*, in this case, are secreted into the bloodstream to be spread throughout the organism's body. Eventually, they are released into the intercellular space and bind to the receptors of receiving cells

Some signal molecules remain bound to the cells that secrete them. Of course, in such cases, the receiving cell has to be in direct contact with the cell sending the signal (Fig. 7.2a). More often, however, the signaling molecules are secreted into the intercellular space and freely diffuse there before reaching the receptor of a target cell (Fig. 7.2b). On the scale of nanometers, diffusion is sufficiently fast, so the signal transmission can occur in the millisecond time range.

In large multicellular organisms, like ourselves, it is often required to transfer signals over distances comparable to the size of their bodies. Two ways of signal transduction over long distances emerged through evolution. The first one uses the circulating blood, where the signal molecules are secreted and delivered to various regions of the body (Fig. 7.2c). In this way, endocrine cells distribute hormones, a special type of signal molecules, through the organism's body.

The delivery of signal molecules through the bloodstream occurs on a timescale of tens of seconds. It is good enough for many needs of the organism, but too slow for the signals causing muscle contraction. This is why multicellular organisms developed the second, very fast, and sophisticated way of signal transduction through the organism's body. In this case, the transduction of the signal occurs in the form of a wave of electrical excitation in special extremely long nerve cells, also called *neurons*. It takes only a few milliseconds for the signal to travel through a neuron from one end of the body to another. We will consider this process in detail at the end of this chapter.

7.3   Intracellular Signaling Molecules

Proteins are the major participants in signal transduction inside the cell. The proteins should be considered as molecular switches that change their conformation between active and inactive states. These conformational changes have to be sustainable and, therefore, they are induced by chemical modifications rather than just temporary interactions with other proteins in the signal transduction chain. Thus, with rare exceptions, the state of a signaling protein depends on its chemical modifications. The most common type of chemical modification is attaching a few phosphate groups to specific amino acids of the protein. The reaction, called phosphorylation, is catalyzed by special enzymes, protein kinases. Such amino acids as serine, threonine, and tyrosine are usual targets for phosphorylation. The phosphorylation of each signaling protein is catalyzed by the specific protein kinase. Therefore, there are hundreds of different protein kinases coded in an animal genome. More often, the signaling proteins are protein kinases themselves, and they form a kinase cascade, when the activation of one kinase phosphorylates the next, and so on. The receptor protein can be the first kinase in the cascade.

If the phosphorylation switches the protein on, the reverse reaction switches it off. The latter switch is equally important since it makes the signal molecule ready to receive the next signal. Therefore, each cell has many enzymes that catalyze the removal of phosphate groups from the signaling proteins, protein phosphatases. The phosphatases have very high specificity to their target proteins, so each phosphatase changes the state of a particular signaling protein.

Instead of chemical modification, the conformation of a protein can be affected by binding a small molecule, and this is also widely used in signal transductions. One such small molecule is guanosine triphosphate (GTP), which activates the signaling protein when it binds with it. The bound GTP can be hydrolyzed to guanosine diphosphate (GDP) and inorganic phosphate, similar to the hydrolysis of ATP molecule (see Fig. 2.7). The hydrolysis inactivates the protein. It can be activated again after releasing the bound GDP molecule and binding another GTP. The signal transduction depends on the special receptor which, when activated, binds the GDP-bound signaling protein and catalyzes the replacement of GDP by GTP. This GDP-GTP exchange is a common step in the signal transduction by the major type of eukaryotic receptors, G-protein-coupled receptors (they are called so because their action depends on binding GDP/GTP), which are considered below.

7.4   G-Protein-Coupled Receptors

G-protein-coupled receptors (GPCRs) regulate an enormous amount of cellular processes. They transfer signals from hormones and neurotransmitters. They are involved in the regulation of metabolism. Our senses of light and smell depend on them. There are more than 1000 different GPCRs coded in the human genome, and each receptor is specific for a particular signal. The signal molecules for GPCRs include peptides, proteins, sugars, lipids, and many molecules that we can taste and smell. Other GPCRs can be activated by light. GPCRs are very important for medicine since the action of nearly half of all drugs is related to the pathways that start from GPCRs.

All GPCRs have similar architecture. Seven α-helices that pass through the membrane form a barrel (Fig. 7.3). The binding site for a signal molecule is located inside the extracellular part of the barrel. The binding of the molecule changes the conformation of the entire protein, including the intracellular part that binds with a specific G-protein. G-proteins are a common part of the signaling pathways associated with GPCRs. The proteins are capable of binding either GTP or GDP molecules that activate or inactivate them. G-proteins are heterotrimeric proteins consisting of α-, β-, and γ-chains. They are always anchored to the membrane where they swim freely. The largest subunit, α, is bound with β- and γ-subunits when it is inactivated (bound with GDP). In this inactive state, the protein can bind to activated GPCR. Each G-protein can bind only its own set of GPCRs. The binding

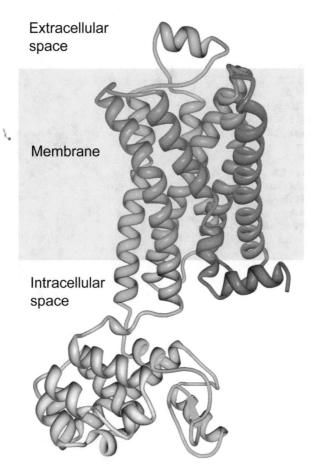

Fig. 7.3 Structure of the G-protein-coupled receptor (Cherezov et al., *Science*, 2007. **318**, 1258–65; PDB ID: 2RH1). Seven α-helices of the protein pass through the cellular membrane (shown in yellow). The image from RCSB PDB was obtained with the Molecular Biology Toolkit (Moreland et al., 2005, BMC Bioinformatics, 6:21)

causes the dissociation of GDP from the α-subunit of the G-protein followed by the binging of GTP from the cytoplasm. This activates the G-protein and it dissociates from GPCR. In its active state, the α-subunit is separated from the β- and γ-subunits (Fig. 7.4). Now the subunits can activate the next signal transmitters.

Fig. 7.4 Activation of a G-protein-coupled receptor (GPCR) and a G-protein, associated with the receptor. (**a**) Both the GPCR and the GDP-bound G-protein are in the inactive state. All three subunits of the G-protein are bound together. (**b**) The GPCR is activated (shown in red) by binding the ligand (green) and binding the G-protein. (**c**) The binding promotes the dissociation of GDP from the G-protein and binding GTP. (**d**) The binding of GTP activates the G-protein, which dissociates from GPCR and is separated into two parts, α and βγ. Both activated parts of the G-protein can now transfer the signal to the other components of the signaling chain

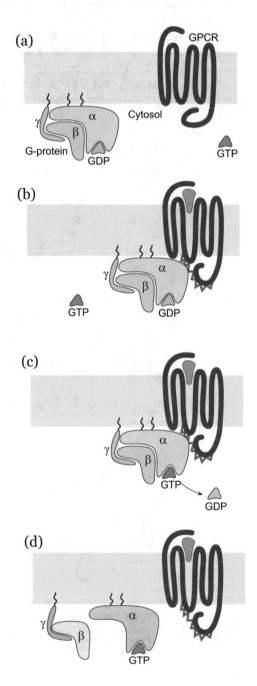

The steps related to the activation of G-protein are always the same, regardless of the specificity of GPCR and G-protein. Further transduction of the signals obtained by GPCRs can vary greatly for different signaling pathways, however. Many G-proteins activate ion channels and proteins participating in transcription regulation. Often G-proteins act through the so-called second messengers, small molecules that transfer initial extracellular signals to various intracellular receivers. Of these second messengers, the most important one is the cyclic AMP (cAMP) shown in Fig. 7.5. This messenger regulates various processes in different types of cells.

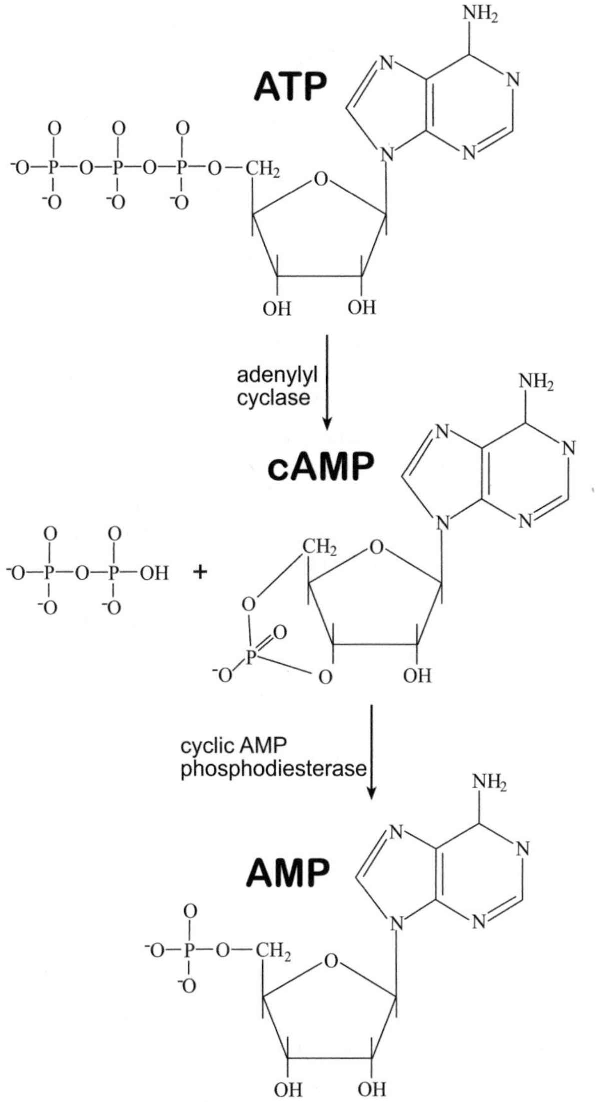

Fig. 7.5 Enzymatic conversion of ATP into cyclic AMP and cyclic AMP into AMP. The majority of carbon and hydrogen atoms are not shown in the structures

In particular, it regulates the production of various hormones in specialized cells. The signals are transmitted by increasing the intracellular concentration of cAMP which can be changed very fast due to the permanent activity of two enzymes. One of them, *adenylyl cyclase*, synthesizes cAMP from ATP, while the other one, *cyclic AMP phosphodiesterase*, continuously converts cAMP to AMP. This allows very fast changes of cAMP inside the cells. Activation of adenylyl cyclase by binding with the activated G-proteins can increase the concentration of cAMP by more than 10 times in seconds. The other extracellular signals, acting through different GPCRs, reduce cAMP levels by activating G-proteins which inhibit the adenylyl cyclase.

The majority of signal pathways mediated by cAMP include the activation of cAMP-dependent protein kinase (PKA). PKA phosphorylates different target proteins in different types of animal cells. In particular, by phosphorylating specific amino acids in the phosphodiesterase, the PKA activates the enzyme which rapidly reduces the cAMP concentration. This negative feedback loop quickly terminates the signal mediated by cAMP.

7.5   Signal Transduction in Neurons

The processes of transmitting signals from one cell to another that have been considered so far are based on the diffusion of small molecules and proteins. For distances of a few μm, the diffusion-based communication is sufficiently fast and well serves the cell needs. However, sometimes the signal has to be transmitted over really large distances, comparable with the size of an animal body, over a very short time. The signals from the brain to the muscle cells are just such an example. Diffusion is well too slow for signal transmission over this scale, and a different mechanism is used in this case. This ingenious mechanism is based on the movement of ions in the electric field.

The transmission of signals over long distances occurs in the nerve cells, neurons. The neurons consist of the cell body, dendrites, and axon (Fig. 7.6). The dendrites are relatively small branches that are connected with other neurons and receive signals from them. The axon can be very long, more than 1 m in length in humans. It has branches at the far end, which connect it with other neurons or with cells of different types. The signals from the body of the neuron travel through the entire length of the axon, and it takes only a few milliseconds! The diffusion of molecules in solution cannot provide such speed. Indeed, the time needed for particles to spread by diffusion is proportional to the square of the distance. Thus, when the distance is changing from the size of a typical animal cell (10 μm) to 1 m, the time of travel increases by a factor of 10^{10}. So, if the signal transmission would be based on diffusion, the time of signal traveling through a long axon would be comparable with the entire lifetime of an animal. Movement in the electrical field could be many orders of magnitude faster, but there is no way to make an electric field of sufficient strength along the entire length of the axon. However, living organisms

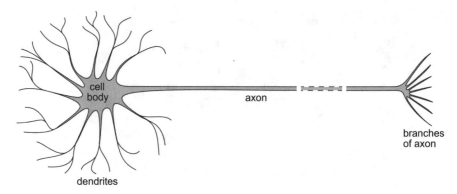

Fig. 7.6 An animal neuron. The nerve signal spreads through the axon from the neuron body to the axon branches at the far end. The length of the axon in animals can exceed 1 m. The dendrites receive signals from other neurons, while the axon branches transmit the signals to other animal cells. There is a high density of Na⁺ and K⁺ channels and transporters in the axon membrane

developed a way of very fast signal transmission along the axons. Regardless of the meaning of the signal, it always spreads like a wave of electric excitation, called the *action potential*. To consider this process in some detail, we have first to discuss the electrical properties of the cell membrane.

7.5.1   Electrical Properties of the Membrane

The cell membrane is nearly impenetrable for inorganic ions, so their concentrations outside and inside the cell can be different. Ion transporters and channels carefully regulate the concentration of various ions inside the cell (see Chap. 3). Since there are many organic negatively charged molecules inside the cell, an excess of small cations is needed to neutralize these negative charges and make the total charge of the cell close to zero. Still, this total charge remains slightly negative.

There are a few types of inorganic ions that are especially important for cell life: K^+, Na^+, Mg^{2+}, and Ca^{2+}. The concentration of K^+ is about ten times higher inside the cell than outside, while the opposite is true for Na^+. The negative charge of the cell causes an increased concentration of positive ions outside the cell membrane, while more negative ions are located in the interior vicinity of the membrane (Fig. 7.7). This ion distribution results in the electrical polarization of the membrane. The electric potential across the membrane of the axon equals −70 mV but can change to +50 mV during the electrical excitation of a membrane spot. These potentials correspond to very large electric fields inside the membrane, up to 100,000 V/cm, since the membrane thickness is close to 5 nm. Such very large electric fields are needed to open and close voltage-gated cation channels in the membrane.

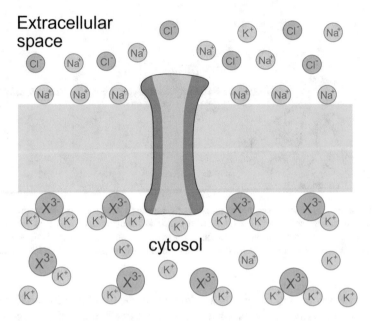

Fig. 7.7 The ionic atmosphere inside (in the cytosol) and outside (in extracellular space) of the cell membrane. The positive ions inside the cell, in the cytoplasm, are mainly represented by potassium ions, while there is a very large excess of sodium ions outside the cell. The negative ions in the cytoplasm are mainly large organic ions of various natures (shown as X^{3-}), often bound with potassium ions. The cell as a whole has a small negative charge, so outside the cell, there are more positive ions in the vicinity of the membrane. Correspondingly, more negative ions are concentrated near the membrane inside the cell. Ion channels and transporters support the distribution of ions. One of many ion-specific channels is shown in the figure

7.5.2   The Action Potential

The signal transduction along the axon of the neuron depends, first of all, on the voltage-gated channels specific for sodium ions. The channels can be in three different states: closed, open, and inactivated (in which the channels are also closed). At the resting state of the membrane, the channels are in the closed state as well. The action potential is initiated by a slight local depolarization of the membrane, caused by a small electric current. This current comes as a signal from another cell. As a result of this depolarization, some of the voltage-gated Na^+ channels open, and Na^+ ions move from outside to inside the cell. This further depolarizes the membrane and opens more Na^+ channels, so the channels work like a system with positive feedback. The flow of sodium ions into the cell increases, since their concentration outside the cell is so much larger than inside. This influx of Na^+ changes the local membrane potential from -70 mV to nearly $+50$ mV (Fig. 7.8). However, the open state of the Na^+ channels is metastable. Under the condition that the membrane is depolarized, the channels pass into the inactivated state where their energy is lower. In this state, they are closed for Na^+. The transition from the open state to the inactivated state takes only about 1 ms. After that transition, the channels cannot be opened again before they return to the closed state. The latter transition occurs only after the spot of the membrane is repolarized again. This occurs because the membrane depolarization opens K$^+$ channels, and K$^+$ ions start flowing from inside to outside the axon spot. Due to the very high concentration of K$^+$ inside the cell, the flow of K$^+$ overpolarizes the membrane (see Fig. 7.8). The Na$^+$ channels, as well as the K$^+$ channels, return to the closed state. Then the ion transporters reduce the concentration of Na$^+$ inside the axon area and increase the concentration of K$^+$. This returns the spot of the membrane and the channels at the spot to the resting state.

The described chain of events, resulting from the small electric current, involves only a tiny spot of the axon membrane. However, the influx of Na$^+$ ions spreads from the initial spot to adjacent areas and results in the initial membrane depolarization there. Since the influx creates a local electric field along the axon, the spreading of Na$^+$ ions represents a fast directed movement rather than diffusion. As a result, the action potential quickly spreads along the axon.

The voltage-gated Na$^+$ channels remain in the inactivated state for a few milliseconds. Over this time, the spot where the excitation occurred at the previous moment does not respond to the spreading of Na$^+$ from the adjacent areas (Fig. 7.9). Due to this property of the channels, the excitation spreads only to the area which was not

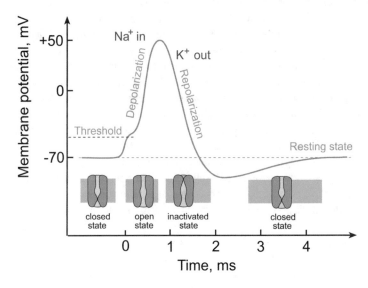

Fig. 7.8 The action potential and Na$^+$ channels. The action potential is initiated by a small electric pulse, which depolarizes a membrane spot above the threshold. This initial depolarization opens the Na$^+$ channels and the influx of Na$^+$ amplifies the membrane depolarization. In about 1 ms after opening, the Na$^+$ channels change their state to the inactivated one. In this state, they are closed for sodium ions but cannot be opened. The depolarization also opens K$^+$ channels and K$^+$ ions flow outside the cell. When the polarization of the membrane spot is restored, the Na$^+$ channels change their state to the closed one, K$^+$ channels close, and the ion transporters reinstall the initial concentrations of Na$^+$ and Na$^+$ inside the cell, in the vicinity of the spot. The states of the Na$^+$ channels are shown under the curve of the membrane potential

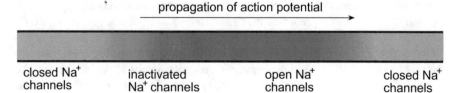

propagation of action potential

closed Na$^+$ inactivated   open Na$^+$ closed Na$^+$
channels   Na$^+$ channels   channels   channels

Fig. 7.9 Movement of the action potential along the axon. Due to the inactivation of Na$^+$ channels by the membrane depolarization, the corresponding spot of the axon membrane (shown by red) cannot fire the next action potential for a few milliseconds. This makes the spread of the action potential unidirectional, creating a wave of electric excitation (it is assumed that the spot of initial excitation is located at the left end of the axon)

involved in the process at the previous moment, creating a fast-moving directed wave of electric excitation.

7.6   Sensing Smell

Detecting smells represents a good illustration of the signaling principles and mechanisms described in this chapter. The process starts from the olfactory GPCR receptors located in the membrane of olfactory neurons in the nose (Fig. 7.10). The receptors can bind, with varying affinities, a range of odor molecules. Correspondingly, a particular odorant molecule may bind many types of olfactory receptors. There are about 400 different genes for the olfactory receptors in humans. However, all olfactory receptors at the surface of each neuron are identical.

The binding of the odorant to the olfactory receptor causes conformational changes in the protein. Due to these changes, the receptor binds and activates the corresponding G protein on the inner side of the membrane. The G protein, in turn, activates the adenylyl cyclase, which catalyzes the conversion of ATP into cAMP. The nucleotide opens cAMP-gated ion channels which allow Ca^{2+} and Na$^+$ to enter the cell, depolarizing the membrane of the neuron and beginning an action potential. The action potential transmits the information along the axon to the brain.

All steps of the signaling pathways responsible for the sensing of smell are identical for all olfactory neurons. The specificity of different odorants exists only at the olfactory receptors. Thus, the brain has to know the correspondence between the neurons and their specific olfactory receptors. Although we do not know the molecular details of the signal treatment by the brain, we know that it handles the job pretty well.

Fig. 7.10 Olfactory
receptor neurons. The
length of the axons shown
here is disproportionately
short. Two neurons are
shown by different colors
to emphasize different
sensory receptors on their
olfactory cilia

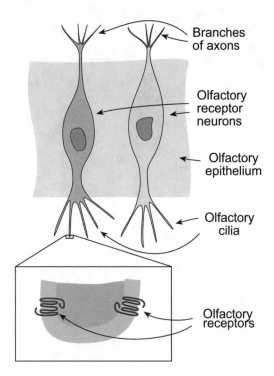

Branches
of axons

Olfactory
receptor
neurons

Olfactory
epithelium

Olfactory
cilia

Olfactory
receptors

Chapter 8
Multicellular Organisms and Their Development

8.1 Body Tissues and Connections Between the Cells

There are four basic types of tissues in the animal body: connective (bone and tendon tissues), muscular, epithelial, and nervous tissue. These tissues consist of different types of specialized cells, and each cell is connected to the other cells or the extracellular matrix. Since the matrix has not been discussed in this book, it is worth describing its properties briefly.

The extracellular matrix represents a network of proteins and polysaccharide filaments secreted by the cells embedded there (Fig. 8.1). The mechanical strength of the tissue is provided, mainly, by special protein collagen, which forms very rigid triple-stranded helices. Depending on the structure of the network, the matrix can be rigid as in bones, or elastic as in tendons. The rigidity of the matrix can be increased enormously due to the binding of the calcium salts by the matrix proteins and polysaccharides, as in structures of bones and teeth. In connective tissues, the cells are isolated from each other, so each cell interacts only with the extracellular matrix. Small molecules, including small proteins, can diffuse through the matrix, so the cells embedded in the matrix can receive all the necessary nutrients. Although the extracellular matrix is relatively dense, the cells are capable of moving and dividing there. To do so, they make an extra space around themselves by digesting nearby parts of the matrix. Digestion is achieved by special enzymes secreted by a moving or dividing cell. The enzymes remain bound to the cell membrane to prevent the digestion of other parts of the matrix. The cells are capable of changing the matrix shape and mechanical properties.

Connective tissue is always covered by an epithelial one, as shown in Fig. 8.2. In epithelial tissue, the cells are connected and form a sheet called epithelia. The cells of the epithelia are also connected with a thin layer of the special extracellular matrix, basal lamina.

The cells of muscular tissues can also be attached to the matrix. The cell-matrix attachments are very important for the development of those cells. Laboratory

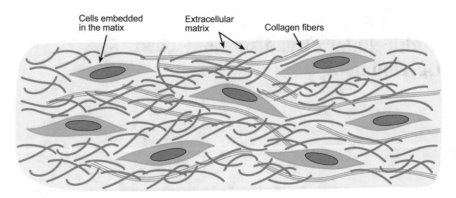

Fig. 8.1 Connective tissue. In the solid connective tissues, bones and teeth, the extracellular matrix is bound with a large amount of calcium salt, which makes up 70% of the tissue mass. The cells constitute a small fraction of the tissue mass

studies show that the corresponding cells are not able to divide if they are not properly attached to the matrix and undergo *apoptosis* (described in detail in the last section of the chapter). When mutations compromise the attachments and their controlling mechanisms in cancer cells, the cells detach from the matrix and spread over the body, causing cancer *metastasis*.

Intercellular connections (or junctions) are a key feature of tissues in multicellular organisms. They are necessary to maintain the organization of the body and to transmit signals between the cells. Typical types of junctions can be illustrated by considering the cells of epithelia (Fig. 8.2). Each type of junction is made of special proteins. Some of the junctions anchor adjacent cells to each other. These junctions are also bound to actin filaments inside the cells, so they are a key element of the continual network of filaments of the tissue (see Fig. 8.2). The network makes an important contribution to the epithelia's mechanical properties. The so-called tight junctions (Fig. 8.3) seal the gaps between the cells, which is very important for the skin epithelia since the sealing prevents the leaking of small molecules across the epithelium. Some junctions form channels for the free flow of small soluble molecules from one cell to another. Other junctions form attachments between the cytoskeleton of the epithelial cells and the basal lamina, called the cell-matrix anchoring junctions.

The cell-cell anchoring junctions are mediated by transmembrane proteins called cadherins. Two identical cadherins can bond with each other (Fig. 8.4). The proteins are extended to the extracellular space, so the membranes of the adjacent cells do not touch one another. The proteins are highly specific for the cell type, and over 180 various cadherins are coded in the human genome. During organism development, some cells switch from the production of one type of cadherin to another, changing the cell-cell interaction.

There are also gap junctions formed by special proteins whose structure reminds the structure of channel proteins (Fig. 8.5). They provide an exchange of inorganic ions and other small molecules between the cells. The exchange is very important

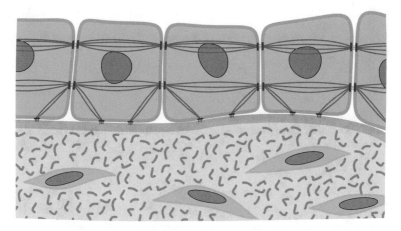

Fig. 8.2 A segment of connective and epithelial tissues. Each cell of the epithelial tissue (shown in green with brown nuclei) is connected to adjacent cells of the same tissue. Epithelial cells (at the top) form the lining for connective tissues (at the bottom). Thin brown lines in the epithelial cells correspond to the microtubule networks

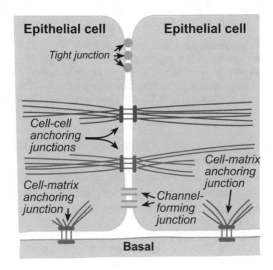

Fig. 8.3 Junctions between the cells of epithelia. There are cell-cell anchoring junctions and tight junctions that seal the gap between the cells. The anchoring junctions are connected to the ends of actin filaments inside the cell. The filaments bind the junctions at the opposite sides of the cell, as shown in Fig. 8.2. There are channel-forming junctions that allow the exchange of small molecules between the cells. The cells are also connected to the basal lamina by the cell-matrix anchoring junctions

for the synchronized regulation of the activity of adjacent cells. The gap junctions connect the interiors of adjacent cells but keep them isolated from the extracellular space. Although the gap junctions are not highly selective, different types of them have a different permeability for small molecules. Similar to the ion channels, the junctions can be in open or closed states, depending on the stimuli they receive from

Cell
membrane

Cell
membrane

Fig. 8.4 Junctions between two cells mediated by cadherins. The proteins usually have five joined identical subunits outside the cells. Identical cadherins exposed on the surface of adjacent cells can bond with each other

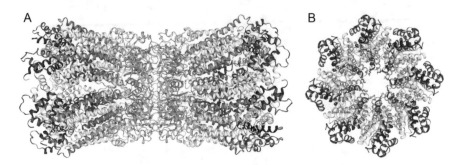

Fig. 8.5 Structure of a gap junction (Oshima et al. *PDB*, 2016; PDB ID: 5H1Q). (**a**) The junction is made of two identical transmembrane proteins, hemichannels, which form a dimer. Each monomer passes through the membrane of its cell, so the dimer connects the interiors of two adjacent cells by a channel, which is well seen in the side view of the protein (**b**). The diameter of the channel is close to 1.5 nm. The images from RCSB PDB were obtained with the Molecular Biology Toolkit (Moreland et al., 2005, BMC Bioinformatics, 6:21)

the cells. Gap junctions are usually clustered in the cell membrane, and each cluster contains thousands of junctions.

Of course, there are junctions between cells of different tissues as well. In particular, the axons of neurons can form junctions with the muscle cells (Fig. 8.6). These junctions communicate signals which cause the contraction of the muscles.

8.2 Development of Multicellular Organisms

The development of extremely complex multicellular organisms from a single fertilized egg is one of the most amazing phenomena of life. Although development is a very tough topic for investigation, one thing facilitates progress in this field. It turned out that the process of development follows the same principles in all multicellular organisms, so we can learn a lot about the development of vertebrates, like ourselves, by studying the simplest animals. Today, many basic principles of development are well understood, although some important questions remain to be answered.

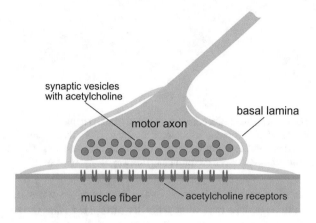

synaptic vesicles
with acetylcholine

basal lamina

motor axon

muscle fiber   acetylcholine receptors

Fig. 8.6 The neuromuscular junction. The axon terminal is covered by a basal lamina. The signal for the muscle contraction is transmitted to the muscle cells through acetylcholine molecules released from the synaptic vesicles. The molecules pass through the basal lamina to reach the acetylcholine receptors

8.2.1   The Basic Mechanism of the Development

Let us first list two evident problems that we can see when we start thinking about development. First of all, one can wonder how a zygote knows about all the structural organization of an adult animal. Of course, we have to accept the fact that all needed information is written in the genome of a fertilized egg. We understand rather well how the protein structure and regulation of their activity are written in the genome. Still, it does not explain incredibly well-coordinated transformations of the embryo in development. So, we need to find what kind of specific information is needed for the process and how this information is written in the genome.

Second, it seems very difficult to imagine that vertebrates with a very complex structure of organs can be developed from a single cell. This is possible, however, because the complexity of the organism can be increased gradually by relatively simple steps. For example, a branching tubular system can be formed gradually by growing from a sphere, as illustrated in Fig. 8.7.

Three major processes contribute to the development:

1. Cells divide, increasing their total number. The process of cell proliferation is rather straightforward, although some cells have to die during development. Cell division also contributes to the process of cell diversification, since daughter cells can be different from one another, as we explain below.
2. Cells diversify and specialize. Cell diversification occurs due to different signals coming from other cells. Changes that occur in the cell depend on these signals and the internal state of this cell. Therefore, the same signal can cause very different changes in different cells. The development of individual cells can cause a dramatic change in the shape and organization of an organ.

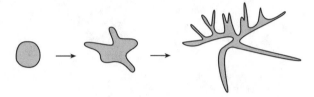

Fig. 8.7 The gradual formation of a complex structure from a simple one. In this illustration, an initial spherical shell of cells forms a branching structure by extending its surface at certain positions. The extension occurs due to regulated cell divisions at these positions. The tubular networks, such as the system of blood vessels, are formed in this way

3. Cells move, changing their location in the organism. As a result, a cell can be located in the organism rather far from its birthplace. The movement of cells makes an important contribution to the process of development.

The development of a multicellular organism is a self-assembly process, so specific interactions between the cells play a key role there. Each cell chooses the way of its behavior depending on the signals it receives from others, first of all, from the adjacent cells. The response of a cell also depends on the signals received by this cell earlier and on the cell ancestors which determine the cell's role in the organism's development. We can say that each cell has a memory that specifies its responses to the new signals from other cells. The previous experience of a cell is registered in the state of its chromatin and the set of transcription regulators that the cell has at the moment.

8.2.2   Differentiation of Cells in the Development of the Organism

All cells of an adult multicellular organism have exactly the same genome, and the difference between them is due to the different expressions of their genes. The differentiation of the cells proceeds gradually as a multistep process. So, the cells become more and more specialized through development, and the set of options available for each of them for further specialization is reducing. Eventually, they reach their final specialization and organize themselves into very complex structures.

Cells specialize due to the signals they receive from other cells. Amazingly, it was found that there are only a few molecules that govern the signaling process during development, and these molecules are nearly the same throughout the entire animal world. This small set of signal molecules is used repeatedly throughout the development of an organism. The same signals cause different results because they interact with various sets of proteins that are expressed in the cell at the moment (Fig. 8.8a). Also, more than one regulator can act on a single cell, and the cell response depends on the combination of the regulators (Fig. 8.8b).

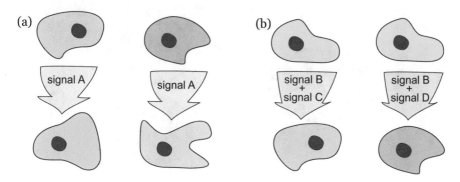

Fig. 8.8 Two mechanisms of generating different cell responses on the signals from other cells. (**a**) The same signal produces different changes in two different cells. (**b**) Different combinations of two simultaneous signals produce different changes in two identical cells

Usually, a cell receives signals from other molecules by binding the secreted signaling molecules with corresponding transmembrane receptors. The binding initiates a chain of conformational and chemical changes in the intracellular proteins, mainly their phosphorylation (see Chap. 7). At the end of the chain is a gene of a transcription regulator or a few genes of regulators that act on different genes. The regulatory region of each eukaryotic gene consists of thousands of nucleotides, so its transcription can be regulated by many proteins in different ways.

Usually, the division of a cell creates two identical daughter cells, but it is not always the case. Under certain conditions, the daughter cells are different, and this makes an important contribution to cell differentiation in early development. Such asymmetric division can occur when some molecules are produced in a certain location of the mother cell, creating a possibility of their unequal distribution inside the cell (Fig. 8.9).

After initial diversification, many more types of cells can be generated by a process called sequential induction. Suppose that two groups of different cells are in contact. The cells of one group can send a signal to the cells of another group, which are in close vicinity. The signaling molecules are transmitted by diffusion, typically at a distance of up to 1 mm. The cells receiving the signal can specialize in a third way (Fig. 8.10), so the total number of cell types increases by one. This way of diversification is used again and again to refine the structure of the developing organism.

8.2.3 Morphogenesis

Differentiation of cells is only one aspect of development. The next aspect is the formation of specific organs, the process called morphogenesis. Not all questions related to this process are sufficiently understood, as we will discuss below.

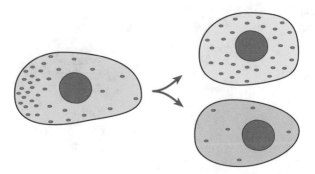

Fig. 8.9 Asymmetric division of a cell during development. The daughter cells are different due to the uneven distribution of certain molecules in the cytoplasm of the mother cell

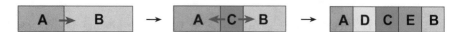

Fig. 8.10 Diagram of cell diversification by the sequential induction. First, cells *A* send a signal (shown by the arrow) to cells *B* which are close to cells *A*. Due to this signal, a part of cells *B* change their type to *C*. Then cells *C* send a signal to the nearest cells *A* and *B* (shown by two arrows). This signal converts parts of cells *A* and *B* into *D* and *E*, correspondingly

Certainly, the migration of cells inside the developing embryo plays an important role there. We consider first how his migration is governed.

It was described at the beginning of this chapter how the cells in an adult body interact with one another. The interaction is complex and very specific. In particular, identical cells expose identical cadherins at their surface, which bind the cells of the same type to one another (see Fig. 8.4). So, to start the migration, a cell has to receive an extracellular signal to eliminate this adhesion. The direction of the movement can be guided in various ways. The cells can move along the network of connective tissues formed earlier in the embryo. This is the case for myoblasts, precursors of the muscle cells. During their trip, the myoblasts permanently test their position by checking adhesion to the surrounding cells. The extracellular matrix of the connective tissues can help cell movement. The end of their trip is marked by the receptors with strong adhesion to the myoblasts. This is where they eventually form the body's muscles. In general, the connective tissues guide not only the cell movement but the shaping of the organs formed by the migrating cells at various locations.

Another way of directing the cell movement is a gradient of an extracellular ligand. The ligand has to be secreted by the cells which are located at the destination site. Of course, chemotaxis requires the ligand receptors at the surface of the moving cell.

The movement of a cell occurs via protrusions on one side of the cell, followed by shrinking on the opposite side (Fig. 8.11). The direction of this movement is specified by the gradient of a substance in the surrounding extracellular space.

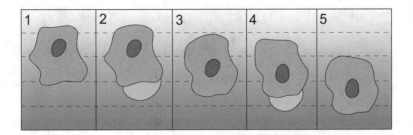

Fig. 8.11 Movement of a cell along the gradient of an extracellular ligand. The ligand concentration increases from top to bottom, and the cell crawls in this direction. A new area (shown by a lighter color) appears on one side of the cell in the region of a higher concentration of the ligand. Simultaneously, the cell is shrinking on the opposite side. Five successive positions of the cell are shown

When a sufficient amount of the needed cells are located at the corresponding site of the developing body, they have to shape the organ. The fate of a cell in this process depends on cadherins exposed on the cell surface. During this self-assembling process, a cell can receive signals from the neighboring cells, which change the exposed cadherins. This changes the specificity of cell interaction with the neighbors. In this way, a simple prototype of the organ can be formed. Gradually, the prototype grows and develops into its final form.

Organs of the animal body consist of many different cells. Correspondingly, the sets of expressed genes are different in these cells. It turns out, however, that regulation of the genes which specify the development of an entire organ can be governed by a single *master transcription regulator*. A master transcription regulator can simultaneously regulate the expression of as many as 50 genes. A striking example of this was obtained in the experiments with the fruit fly Drosophila, one of the favorite objects in the studies of animal development. The formation of eyes in the fly is governed by the master transcription regulator called eyeless. When the researchers artificially expressed eyeless in a group of cells, which were the precursors of the legs, developed eyes were formed on the legs.

Interaction between the cells of a developing organ and the surrounding tissue plays an important role in morphogenesis. The branching of the bronchial tree in the developing vertebrate lung gives a good example of such a process. The tree has to be highly branched to increase the surface of the airways, which deliver oxygen to the lung. At least two ligands participate in the regulation of the branching of the growing tree. The process starts from a bud. The cells of the surrounding tissue located in the closest vicinity to a bud of the growing tree secrete ligand A (Fig. 8.12). This ligand stimulates the extension of the bud. The tip of the bud, on the other hand, secretes ligand B to inhibit the production of A near the tip of the bud. This results in the branching of the growing bud.

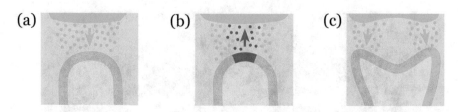

Fig. 8.12 Branching of the bronchial tree. (**a**) Ligand A is secreted by the embryonic cells (green) surrounding a bud of the growing tree. The ligand stimulates the bud extension in the direction of the ligand. (**b**) The epithelial cells (light brown) at the tip of the growing branch secrete ligand B (red), which inhibits the production of ligand A by the nearest cells. (**c**) As a result, the two centers producing ligand A appear. Now ligand A stimulates the formation of a new branch point

8.2.4 Sizing and Timing

Overall, the information related to morphogenesis is written in DNA in very complex and sophisticated signaling networks that regulate the transcription regulators. Numerous negative and positive feedback loops provide the necessary interaction between the cells of the developing organism. Although not all details of the process are known, the general principles of its steps seem to be understood now. Two major problems in this process deserve separate attention, however.

The first problem is the sizing of organs and the entire organism. Somehow the organism has to measure the size of the developing organs to send signals to halt their growth when they reach certain sizes. Although there is a good understanding of the signals which regulate the growth of the organs, it remains a mystery how the moments for these signals are determined or how the growing organism assesses its size.

In general, the size of the organ depends on the number of cells and on the size (volume) of individual cells (it also depends on the volume of the extracellular tissue, but we will ignore it now). So, what kind of information is the organism detecting, the total number of cells or the size of an organ? It turned out that the answer depends on the organism. In plants, more often, the total number of cells in the organ is kept constant. When the size of the cells is artificially doubled, the organ size is doubled as well. However, mammalians monitor the size of the organ rather than the total number of cells in this organ. There is a way to reduce or increase the number of cells in some animals, and the corresponding experiments have shown that changing the number of cells does not affect the animal size.

There are signals sent by special regulators called hormones, which affect the growth of the entire organism. They are usually secreted by a special gland into the bloodstream and spread through the entire body. Some hormones affect the activity of hundreds of genes. The growth hormone is an example of such regulator. Its excessive production causes an increase in the animal size above the normal, and reduction of the hormone results in dwarfism. The hormones help to synchronize the growth of individual parts of the developing organism which is critically important. If all processes are well synchronized, it may be enough to monitor the progress of just a few of them to assess the size of the growing body. Still, specific ways to address the size of the body remain to be discovered.

8.3   Stem Cells

The goal of the development is a well-functioning adult organism capable of reproduction. Still, it does not mean that when the goal is achieved, the production of new cells stops. The cells in the majority of tissues of adult organisms are in the permanent process of renewal when old cells die and are replaced by newly born ones. Somatic cells of the human body can live from a couple of days to a few years. Renewal is necessary for the repair and regeneration of tissues and organs. However, the adult organism consists mainly of fully differentiated cells that are not capable of dividing. The renewal is possible due to the stem cells, which are always present in the adult body. The stem cells are not completely differentiated and can produce various types of fully differentiated cells. Stem cells have attracted a lot of attention over recent years due to the hope of using them for the regeneration of various parts of the human body. Below we briefly consider the major properties of stem cells to give an idea of the cell renewal process.

8.3.1   General Properties of the Stem Cells

There are many types of stem cells, and the cells of each type are responsible for the renewal of a particular tissue in the mammalian body. The stem cells are not universal, so the division of the stem cells of each particular type can only produce fully differentiated cells of certain types. Thus, different stem cells provide regeneration of the skin tissue, lining of the gut, various connective tissues, muscles, blood cells, etc. A dividing stem cell is capable of producing several different types of differentiated cells as well as stem cells of the same type. The fate of the daughter cells depends on signals from the local environment. On average, each of the daughter cells has a 50% chance of becoming a stem cell and a 50% chance to get on the pathway of terminal differentiation. Thus, the total amount of stem cells of any particular type is maintained over the life span of the adult organism. Usually, the stem cells divide slowly, and to accelerate the renewal of differentiated cells, the pathway from the stem cells to differentiated cells involves the formation of the transit-amplifying cells (Fig. 8.13). The latter cells are still *multipotent*. They go through a certain number of rapid divisions before proceeding to terminal differentiation. The fact that the stem cells do not go through many rounds of divisions reduces the probability of mutation in their genome, which is very important since they have to function properly throughout the lifetime of the organism. The transit-amplifying cells, on the other hand, can only go through a limited number of divisions, so mutations in their DNA are not so critical. At some moments of life, there

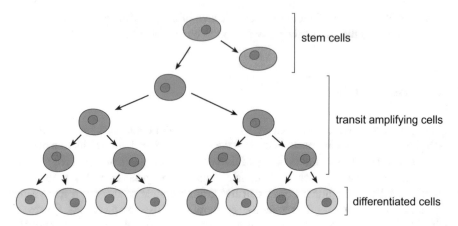

Fig. 8.13 The pathway from the stem cells to the terminally differentiated cells. Dividing stem cells can either replicate themselves or produce the transit-amplifying cells. The latter cells rapidly go through a few rounds of division and then produce various differentiated cells that are not capable of division

is a need for a larger amount of stem cells in a particular tissue, and the organism responds properly.

Although usually tissue renewal starts from the stem cells, there are some exceptions. The fully differentiated insulin-producing cells in the pancreas are capable of dividing and renewing their population themselves. Similarly, the liver cells can divide and can greatly increase the division rate when the need comes. On the other hand, some tissues such as the sensory hair cells in the ear and the photoreceptors in the eye cannot be renewed over adult life. When some of these cells are lost, they are lost forever. This is why vision and hearing decline in old age.

8.3.2   Blood Regeneration

We now consider the regeneration of blood cells as an example of tissue renewal. Blood consists of red blood cells, or erythrocytes, and white blood cells, or leukocytes. Erythrocytes deliver oxygen to nearly all tissues of the body and take out carbon dioxide. White blood cells are responsible for the immune response of the organism. They include *lymphocytes*, *granulocytes*, and *macrophages*. In Chap. 10, we will consider the cells of the immune system in detail. Blood circulation through the system of blood vessels allows fast delivery of the needed cells to any tissue of the body. The network of blood vessels is capable of fast remodeling and adjustment to deliver an increased amount of blood cells to the places where they are especially needed, repair injuries, and fight infections. While the erythrocytes always remain in the blood vessels, the leukocytes use the blood circulation only as a transportation system to be delivered to the needed location where they migrate out from the blood vessels.

Like many other cells of the body, blood cells have a limited life span, about 4 months on average, and therefore are in the process of constant renewal. All different cells of the blood are produced from the common stem cells, located in the bone marrow. The total amount of blood stem cells is relatively small, so their progenitors (transit-amplifying cells) have to pass through multiple divisions to produce a sufficient amount of fully differentiated cells (Fig. 8.14). The progenitors divide rapidly, but the number of their divisions is limited. After these divisions, the cells terminally differentiate. However, some of the fully differentiated leukocytes can divide, which is especially important for *B lymphocytes* (*B cells*).

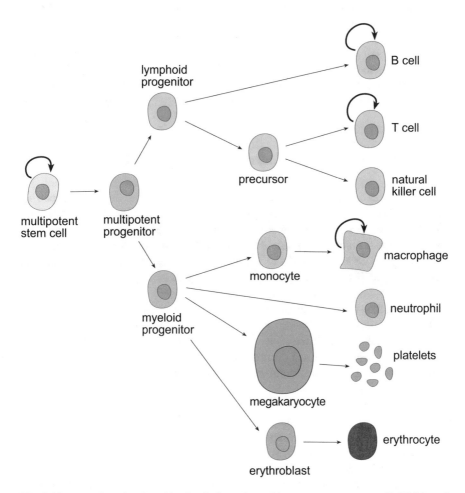

Fig. 8.14 Formation of various blood cells from the multipotent common stem cell. Division of the multipotent stem cell can form more stem cells or multipotent progenitors which are capable of fast division. The progenitors and precursors eventually form fully differentiated red and white blood cells. B cells, T cells, and macrophages are capable of dividing to multiply their number. The figure presents a simplified diagram of the process

The erythrocytes constitute more than 99% of the total number of blood cells. At the end of their development, the cells are densely packed with hemoglobin and contain no usual cell organelles, including nuclei. All the organelles were extruded from the cells during the development. So, the erythrocytes can be considered as bags with protein hemoglobin, which binds to O_2 and CO_2. Clearly, the erythrocytes cannot divide. In humans, they live about 4 months before being killed and digested by macrophages.

The leukocytes, the opposite of erythrocytes, consist of a few classes of cells. B cells make antibodies, a key component of the adaptive immune system. T cells are also a part of the adaptive immune system; they kill cells infected by viruses and bacteria and activate B cells. Macrophages and neutrophils phagocytose and digest invading bacteria and damaged senescent cells. Natural killer cells kill viruses as well as some cancer cells. Platelets, which are produced by megakaryocytes, represent cell fragments rather than entire cells. They initiate blood clotting. There are also some other types of leukocytes. All this variety of white blood cells, as well as red blood cells, originates from the multipotent stem cells located in the bone marrow.

8.3.3 Embryonic and Induced Pluripotent Stem Cells

Due to the stem cells, animals are capable not only of renewing tissues but repairing them after injuries as well. The ability of tissue repair is amazing among some very simple animals. A small freshwater worm can regenerate the entire adult body from its small part. A much more complex animal, a newt, is capable of regenerating an entire organ. The ability of tissue regeneration, however, is much more modest among mammalians. In humans, in particular, only a small part of the neural cells in the brain are normally regenerated. If other nerve cells are lost due to a sickness or an injury, they are lost forever. Still, the examples from simple animals give hope that a wider spectrum of tissues can be regenerated with some artificial help. This hope stimulated great interest in stem cell research over the last decades. Experimental results obtained on mammalians support this hope. The researchers learned how to grow mouse neural stem cells in a culture and then implant them in a mouse brain. These stem cells developed into fully functional neurons in their new environment.

The possibility of artificial help with tissue regeneration requires in vitro manipulation with stem cells. The researchers have learned to grow them in a culture without differentiation or, by changing the growing conditions, to differentiate into cells of specific types. One limitation of this approach to tissue regeneration is the specialization of the stem cells in the adult body of mammalians. Normal multipotent stem cells maintain this specialization, so the stem cell of a particular type is capable of producing the progeny of differentiated cells of only one tissue. It would be very helpful for research and medical applications to overcome this limitation of specialized stem cells.

We know that a fertilized egg produces all cells of the adult body. From the point of cell specialization, the egg represents a *totipotent cell*. This cell is not a stem cell, however, since it cannot reproduce itself. Instead, its successive divisions make more and more specialized cells. Nevertheless, it turned out possible to take cells of the early embryo and grow in culture a line of cells that can divide indefinitely, reproducing themselves. The cells of this line are capable of producing nearly all types of cells when they are placed in a proper environment. This kind of cells are said to be *pluripotent*. They certainly fit the definition of stem cells and are called *embryonic stem cells* (ES cells). The cells are very useful for the study of cell specialization. They made it possible to switch between cell culture, where various genetic manipulations and selections are much easier, and living organisms. Clearly, this opens enormous opportunities. However, the way of obtaining ES cells runs through evident ethical complications.

It has been known for years that cell specialization is reversible. There were successful experiments where the nucleus from a specialized cell replaced the nucleus of a fertilized egg, and an adult animal grew from the egg. This means that a set of transcription regulators, DNA methylases, and histone modifiers, which are present in the cytoplasm of the egg, are eventually capable of reprogramming the cell chromosomes after the transplantation. So, maybe reprogramming could be performed in cell culture, so a culture of regular stem cells or differentiated cells from an adult body could be transformed into pluripotent stem cells. A long search for genes whose expression is critical for pluripotent cells brought eventually the desired result. It was found that a specific set of four transcription regulators added to a culture of fibroblasts is capable of transforming them into *induced pluripotent stem cells* (iPS cells). Under proper conditions in the cultures, these cells can be transformed into nearly any type of specialized stem cell. Like ES cells, iPS cells can divide indefinitely in the cell culture. Thus, the properties of iPS cells are very close to ES cells. By adding proper transcription regulators, the culture of iPS cells can be transformed into desired differentiated cells. Today, iPS cells can be obtained from many types of differentiated cells. Clearly, this great achievement opens remarkable opportunities in the field. iPS cells can be used for tissue regeneration by transplanting the specialized stem cells into an individual for tissue regeneration. The experiments on animal models have brought very promising results there. Another direction of the medical application of iPS cells is correcting disease-causing mutations in individuals (Fig. 8.15).

8.4 Death of the Cells

Cells of the multicellular organism divide, grow, and die. Death of cells in living organisms can occur by *apoptosis* and by *necrosis*. Apoptosis, or programmed cell death, is very important for the organism, and tens of billions of cells in the human body die through apoptosis every day. Apoptosis is a carefully regulated and

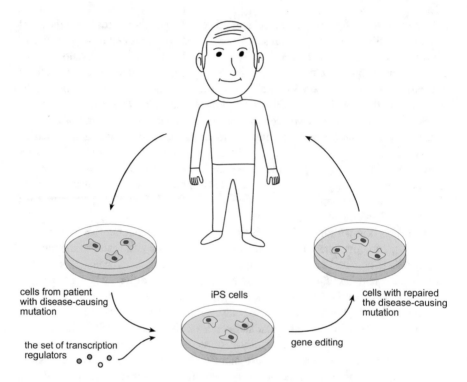

Fig. 8.15 A possible way of using iPS cells to repair disease-causing mutations. Differentiated cells from a patient with a disease-causing mutation are grown in the culture and converted into iPS cells by adding the set of transcription regulators. Then the mutation is corrected by the gene-editing procedure. After testing the result of the editing, the cells with the repaired mutation are differentiated in the desired way and transplanted to the patient. Since the procedure involves only the patient's cells, there is no danger of the immune rejection of the repaired cells

precisely elaborated process. In the course of this process, the dying cell systematically destroys itself from the inside. The remains of the dying cell are completely eaten by other cells, so no trace of it leaves. This is contrary to necrosis, a cell death caused by injury of insufficient blood supply. Necrotic cells spill their remains to neighboring cells coursing the inflammatory response in the surrounding tissue.

Apoptosis is critical for the organism over its entire life. In development, the morphology of the body is changing, and some parts of the body have to be eliminated through apoptosis of the corresponding cells. In the adult organism, many tissues are permanently renewing, so while new cells are produced, others have to die. Cells die through apoptosis when they become infected or deviate from normal development. In particular, cells can detect DNA damage and, if the damage is not repairable, they die by apoptosis. Below we consider the process and its regulation in some detail.

8.4.1   Basic Steps of Apoptosis

Each mammalian cell has special proteases that digest thousands of different proteins during apoptosis, destroying the cell. These proteases, called *caspases*, are not activated during normal cell life. There are two classes of caspases, initiator caspases and executioner caspases. The major role of the initiator caspases is to activate the executioner caspases. First, when a cell receives an apoptotic signal, inactive initiator caspases activate themselves. They bind with an adaptor protein, which stimulates the dimerization of the inactive caspases (Fig. 8.16). In the dimer, each caspase cleaves a designated peptide bond of its partner to stabilize the active form of the enzymes. The initiator caspases activate the executioner caspases by cleaving a certain bond of their backbone. At this point, the process becomes irreversible since activated executioner caspases cleave peptide bonds of numerous proteins in the dying cell. Among those are proteins that keep specific nucleases in an inactive form, so their destruction activates the nucleases. The active nucleases degrade DNA in the cell nucleus. Among other targets of the caspases are cell-cell adhesion proteins that connect the apoptotic cell to its neighbors. After the cleavage of these proteins, the dying cell shrinks, and its cytoskeleton collapses. The chromatin of the cell condenses and splits into fragments. If the cell is large, it breaks into a few fragments enclosed by the membrane. The surface of the membrane is changed chemically, so it can be easily recognized by the neighboring cells or macrophages (large cells dedicated to digesting dying cells and microbes). The cell disappears without any traces.

8.4.2   Apoptotic Signaling Chains

Apoptosis is very carefully regulated to avoid disastrous consequences for the organism. There are two major mechanisms of apoptosis activation called extrinsic and intrinsic pathways. Over the extrinsic pathway, the cell receives apoptotic signals from other cells, while in the intrinsic pathway, the signal is developed inside the cell. We will consider these pathways separately.

The majority of cells have *death receptors* on their surface. The receptors are transmembrane proteins that transmit the apoptotic signals to the cell that has to undergo apoptosis, from surrounding cells. The receptors are activated by binding the ligands which appear on the surface of a neighboring cell, which generates an extrinsic apoptotic signal (Fig. 8.17). The activation is achieved through a conformational change in the receptor, which involves its intracellular domain. The conformational change in the domain increases its affinity to the adaptor protein which binds the receptor. The complex then binds the initiator caspases and activates them, starting apoptosis.

A simplified picture of the intrinsic pathway of apoptosis is shown in Fig. 8.18. Each cell has a few pro- and antiapoptotic proteins. Under normal conditions, they form heterodimers and inactivate each other. When an apoptotic stimulus increases the concentration of inhibitors of the antiapoptotic proteins, the inhibitors replace proapoptotic proteins in the dimers. Released proapoptotic proteins bind to the outer membrane of the mitochondria. This binding somehow releases intermembrane mitochondrial proteins into the cytoplasm. It turned out that these proteins, first of all cytochrome c, play a key role in the pathway. They bind the adaptor protein and activate it. The active adaptor proteins activate the initiator caspases, as shown in Fig. 8.16.

The apoptosis pathways are regulated by changing the concentrations of participating proteins. In many steps of the pathways, the interactions between the proteins occur only in relatively large complexes where each protein presents in a few copies (not shown in Figs. 8.16, 8.17, and 8.18). As a result, the concentration of the active complexes strongly depends on the concentration of their components. Consequently, the sensitivity of the pathways to the concentration of the participating proteins increases.

The apoptotic pathways are activated by apoptotic stimuli. These stimuli can have different origins. In the case of the intrinsic pathway, they appear as a response to cell stress. In one case of stress, when DNA is severely damaged and cannot be repaired, the concentration of the tumor suppressor protein p53 sharply increases. The protein stimulates transcription of the gene, which codes the inhibitor of the antiapoptotic protein, triggering the cascade shown in Fig. 8.18. Due to this role in the intrinsic pathway of apoptosis, the normal functioning of p53 is so critical for the defense of the organism against cancer (see Chap. 12).

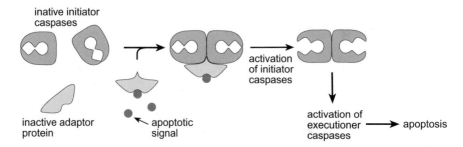

Fig. 8.16 Activation of apoptosis. At normal conditions, the adaptor protein is inactive. It is activated by an apoptotic signal which changes the protein conformation. Active adaptor protein promotes the dimerization of the initiator caspases. In the dimeric state, the caspases activate each other. The active initiator caspases cleave a specific bond in the executioner caspases to activate them. The executioner caspases cleave many other proteins of the cell, starting its elimination

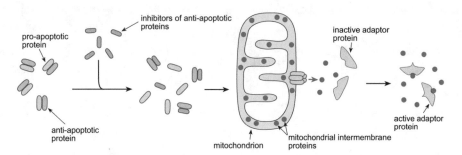

Fig. 8.18 The intrinsic pathway of apoptosis. Each cell contains approximately equal amounts of pro- and antiapoptotic proteins (pink and blue) that bind and inactivate each other. Apoptosis starts when the concentration of the inhibitors of the antiapoptotic proteins sharply increases. The inhibitors replace proapoptotic proteins in the dimers. The released proapoptotic proteins bind with mitochondria and release mitochondrion intermembrane proteins into the cytoplasm. These proteins bind with the adaptor proteins and activate them

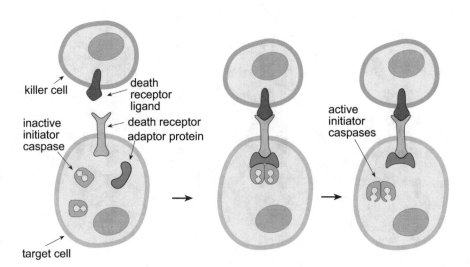

Fig. 8.17 The extrinsic pathway of apoptosis. The apoptosis signal comes from a death receptor ligand located on the surface of a neighboring cell ("killer cell" in the figure). Binding the ligand changes the conformation of the death receptor, so the adaptor protein located inside the target cell can now bind the death receptor. Consequently, the adaptor protein changes its conformation and becomes capable of binding to a pair of initiator caspases. Binding with the adaptor protein causes the dimerization of the caspases and activates their protease activity. Each caspase cuts a specific peptide bond in its partner, making both of them capable of activating the executioner caspases

Chapter 9
Biological Defense I: Protection from Foreign Nucleic Acids

9.1 General Remarks

Every living organism has a defense system protecting it from invasion. Usually, features of more complex organisms can be considered as a development of the corresponding features of more simple organisms. It is not the case with the defense systems. There are several types of defense systems that use completely different principles. Considering these systems, we cannot avoid thinking that many more defense systems could be designed. The review of these systems strongly suggests that the way of life on our planet is just one particular way among many other possibilities.

   An ideal defense system should be able to destroy any invader but preserve all own structures of the organism. This is a very difficult requirement since the invaders consist of the same materials, nucleic acids and proteins. Somehow the proteins or nucleic acids of the invader have to be distinguished from the organism's own macromolecules and recognized as foreign. Despite all sophistication of the existing defense systems, none of them satisfies this requirement to the full extent. Still, more complex systems of vertebrates perform their work much better than the simplest bacterial defense systems. In this chapter, we briefly review the simplest systems based on the recognition of certain sequences of nucleic acids, constituting the genomes of invading pathogens. The most sophisticated defense systems of vertebrates will be considered in Chap. 10.

9.2 The Restriction-Modification System

This system is used in bacteria. It is based on the chemical labeling of bacteria's DNA to make it distinguishable from the DNA of invading bacteriophages (viruses of bacteria). The labeling allows special bacterial enzymes to digest the foreign DNA molecules which remain unlabeled. Let us consider the system in some detail.

A. Vologodskii, *The Basics of Molecular Biology*,
https://doi.org/10.1007/978-3-031-19404-7_9

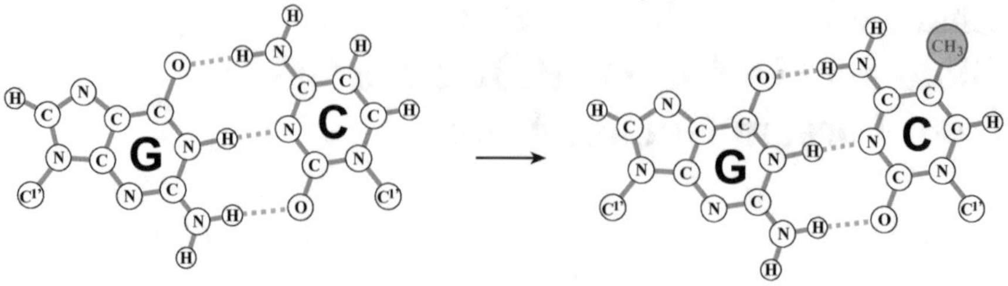

Fig. 9.1 GC base pair with unmethylated and methylated cytosine. The hydrogen bonds between the bases are shown by dashed lines. The newly added methyl group is shaded. The methylation does not affect the base pairing and does not disturb the incorporation of the base pair into the double helix but can be easily recognized by DNA binding proteins

To label their DNA, bacteria methylate cytosines at segments with specific sequences. The methylation does not affect the base pairing and DNA replication, although it can change the interaction between the methylated DNA segments and proteins that bind to these segments (Fig. 9.1). DNA methylation is widely used in biology, mainly for different types of regulation of gene expression (see Sect. 5.1.3). Special enzymes, methyltransferases, perform the methylation. The enzymes methylate cytosines which are located at certain positions of short DNA segments with specific sequences 4–6 bp in length. The sequences of these segments are palindromic, so they are identical in both complementary strands. The cytosines in both strands of the segments are methylated. The methyltransferases of different bacteria recognize and methylate different specific sites. In each bacterium, these enzymes are complemented by the restriction nucleases that recognize the same sequences, but only if cytosines in the recognition sites are not methylated. The nucleases bind to these DNA segments and cut both strands at specific dinucleotide steps (see Fig. 4.12). So, unmethylated DNA molecules are split into fragments and cannot serve for reproduction.

Two conditions have to be satisfied to make this system effective. First, the methylation of the host DNA has to be reproduced in the replicated molecules during cell division. It is achieved due to the properties of the methyltransferases that bind mainly with the DNA segments that already have methylated cytosine in one strand. This methylated strand comes from parental DNA that was methylated in both strands (Fig. 9.2). The enzymes bind to the semi-methylated segments and methylate the cytosine in the newly synthesized strand. This preserves the methylation of the host DNA during cell division. Second, the foreign DNA has to be cut before it gets methylated. This condition is usually satisfied since methylation of fully unmethylated specific sites of the foreign DNA is a slow process. So, the restriction nucleases manage to cut these DNA molecules at least partially before they become methylated.

Although around 25% of known bacteria use the restriction-modification defense system, it has an evident weakness. If the methyltransferases miss methylating a particular site of the bacterial DNA before the next round of DNA replication, both

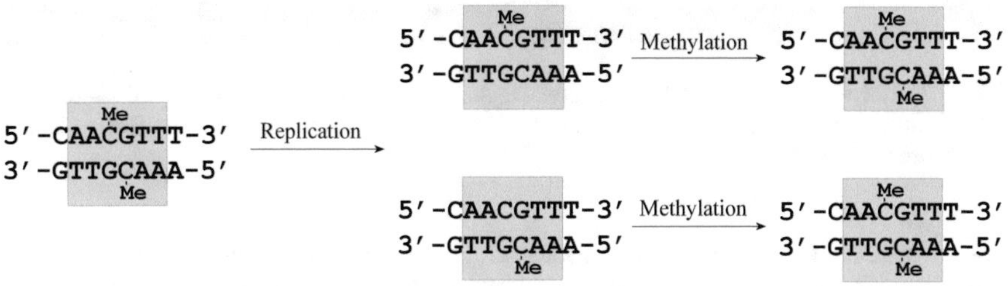

Fig. 9.2 Preserving the methylation of specific sites during DNA replication. Although the cytosines in the newly synthesized strands are not methylated after DNA replication, the methyltransferases, guided by the methylated cytosines in the parental strand, quickly complete the modification

strands of one of the daughter double-stranded DNA molecules will be unmethylated. The restriction enzyme will cut this site.

Remarkably, the sequence-specific restriction nucleases that are a part of this defense system became the basis of genetic engineering, which revolutionized the research in molecular biology (see Sect. 4.6.1).

9.3 CRISPR-Cas System

CRISPR-Cas system was discovered only recently, in the first decade of the twenty-first century. It is now known, however, that nearly all archaea and around half of bacteria have CRISPR-Cas systems. The system is one of a few that are based on short RNA molecules that have homology with the invading foreign DNA. The sequences of these RNAs are coded in a locus of the genomic DNA called the clustered regularly interspaced short palindromic repeats, or CRISPR. CRISPR loci consist of several repeats separated by segments of variable sequences called spacers (Fig. 9.3). The spacers correspond to segments of viral or plasmid DNA captured by previous generations of the bacterium during the invasions of these pathogens. The lengths of the repeats and spacers depend on a particular CRISPR system but are in the range of 20–50 and 20–70 bp, respectively. With the help of special Cas proteins, the short RNA molecules transcribed from the spacers, crRNAs, can replace one of the DNA strands in the corresponding segment of the foreign DNA molecules. This results in the formation of the hybrid RNA/DNA duplex. Then the DNA segment bound with crRNA is digested by special nucleases (Fig. 9.4). Thus, the CRISPR-Cas system defends against foreign DNA molecules that have homology with one of the spacers of the CRISPR locus. Typically, the locus contains not more than 50 spacers, and each spacer provides defense from only one type of foreign DNA.

The set of spacers of the CRISPR locus is changing as a response to the viral/plasmid invasion. Somehow the system is capable of adding a new

| cas genes | | | | | | | | | | | | | |

Fig. 9.3 The structure of CRISPR locus of DNA. The gray rectangles correspond to identical palindromic repeats. The spacers which are complementary to segments of various foreign DNAs are shown in different colors

repeat-spacer unit to the CRISPR locus with a spacer from the invading DNA. The new unit is added at the leader end of CRISPR. Since the growth of CRISPR due to such additions has to be limited, the older units at the opposite end of the locus are deleted. Thus, the CRISPR locus of a cell contains information about the history of attacks on the cell progenitors by foreign DNA molecules.

Many proteins are involved in the functioning of this defense system. A part of them is coded in the DNA locus adjacent to CRISPR, while other proteins are distributed along the genome. Many mechanical details of the system functioning remain unknown.

A special mechanism is used by this system to avoid the destruction of the bacteria's own CRISPR locus. The segments of a foreign DNA which can be recruited to CRISPR as spacers, selected by the protein machinery, must have a specific sequence of three to four nucleotides immediately adjacent to the proto-spacers, called PAM (proto-spacer adjacent motif) (see Fig. 9.4b). The presence of this motif is necessary for the interaction with Cas proteins that digest the invading DNA. Thus, the complex of Cas proteins and crRNA interacts with foreign DNA by forming the DNA/RNA duplex and also by direct DNA-protein interaction with PAM. The absence of the PAM sequence in the CRISPR locus itself prevents the digestion of the locus by the same complex, so it precludes an "autoimmune" response.

It is not easy for the CRISPR system to distinguish between foreign and bacterial DNA when it selects proto-spacers. Clearly, the acquisition of a spacer from the cell's own DNA would result in self-destruction. Recent data show that such acquisitions happen, although not very often.

CRISPR-Cas is an adaptive immune system since the CRISPR locus can change as a response to new external conditions, like the appearance of a new virus. Although the available data suggest that the inclusion of a new spacer into the CRISPR locus takes many cell generations, it eventually happens. On the other hand, CRISPR is a part of the cell genome, so it is inherited. Thus, the system shows that the Lamarckian evolution theory, which postulates that adaptation to external conditions can be inherited, turned out to be not completely wrong.

The application of the CRISPR-Cas system to genome editing is discussed in Sect. 4.6.2.

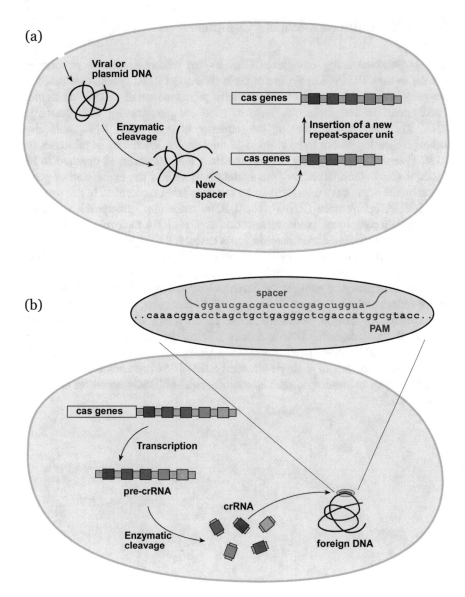

Fig. 9.4 CRISPR-Cas defense system of the bacterium cell. (**a**) Adding a new spacer to the CRISPR locus (shown in red) as a response to a phage/plasmid invasion. A proto-spacer is cut by the bacterial proteins from the foreign DNA and inserted at the leader position of the locus. (**b**) The pre-crRNA is transcribed from CRISPR and cut into crRNAs, each carrying a transcript of a single spacer and a part of the repeat. With the assistance of Cas proteins, crRNA can form a complex with a complementary single-stranded segment of the foreign DNA (the second strand of the foreign DNA and the complex of Cas proteins are not shown). The complex formation requires the presence of a specific PAM sequence in the foreign DNA (shown in green in the extension) that interacts directly with the protein complex. The complex formation is followed by the digestion of this DNA

9.4   RNA-Induced Silencing Complexes

RNA-induced silencing complexes (RISCs) are widely spread in eukaryotes. Similar to the CRISPR-Cas system of prokaryotes, the RISC systems are based on the recognition of target RNA by specially processed small ssRNA molecules. These small ssRNA molecules have to have complementarity with the target RNA. There is no restriction on the sequence of these small RNA molecules, although their length has to be close to 22 nucleotides. Therefore, in principle, the RISC system can silence any single-stranded RNA. However, as opposed to the CRISPR-Cas system, RISC systems mainly only reduce the expression of genes coded by the organism's own DNA or by invading DNA molecules. In this process, called RNA interference (RNAi), small RNA molecules, incorporated into the multi-protein complexes, guide the selection of target RNA molecules whose activity has to be suppressed. Although in some cases, RISC digests the target single-stranded mRNA molecules, they mainly only inhibit their translation. Thus, RISC systems reduce or silence the activity of specific genes that can have either internal or external origins. According to the current view, the major task of RISC is silencing or reducing the activity of internal genes, in particular, transposons (see Sect. 4.3). Clearly, for the goal of defense, RISC cannot attack DNA since it would kill the cell, although digesting DNA is a more efficient defense strategy against invading viruses. Still, in many cases, the defense role of RNAi is very important. If the viral genome consists of a single-stranded RNA, RISC systems can destroy the genome of such an invading virus. Three known types of RISC systems are described in Sect. 5.3.

Chapter 10
Biological Defense II: The Immune Systems of Vertebrates

10.1 The Adaptive Immune System

In general, the full-scale response of the adaptive immune system takes a few days. Over this time, the pathogen can proliferate enormously and do a lot of harm to the body. However, the adaptive system has a memory of previous infections which allows for a rapid response. If the same pathogen invades the organism after it was once erased by the adaptive immune system, the system responds much faster than it did for the first time. This ability of a fast response to previous infections, called immunity, is extremely important for organisms. The memory about a particular infection can be also created by a vaccine that imitates the actual pathogen. Vaccination has become a very efficient tool in fighting various infections. Mechanisms of this memory will be clarified later when we consider the adaptive immune system in detail.

10.1.1 B Cells

The adaptive immune system of vertebrates is based on specialized cells, which are called lymphocytes. There are two major types of lymphocytes, *B cells* and *T cells*. Each organism has billions of different families, or clones, of B cells, and each clone carries unique B cell receptors (BCRs) at the cell surface. The receptors bind antigens—viruses, microbes, and large foreign molecules. They are capable of binding antigens with remarkable specificity and very high affinity, and the binding eventually results in the destruction of the pathogen. Many thousands of BCRs that are exposed on the surface of each particular B cell are identical, and each of them has two identical antigen-binding sites.

The huge diversity of the BCRs preexists regardless of exposure to antigens. Therefore, each organism always has clones of B cells with BCRs capable of binding any pathogen. The binding of the pathogen and interaction with an activated

A. Vologodskii, *The Basics of Molecular Biology*,
https://doi.org/10.1007/978-3-031-19404-7_10

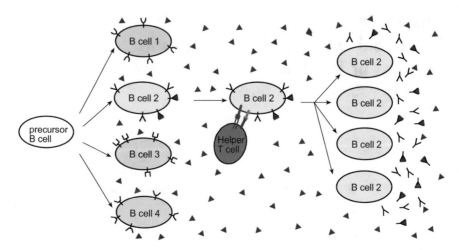

Fig. 10.1 Development of B cells. There are billions of B cell clones with different receptors exposed on their surfaces, but each B cell produces receptors with identical antigen-binding sites (for simplicity, only one of two binding sites of each receptor is shown). When the membrane-bound receptors bind antigens (shown in red triangles), and the B cell interacts with an activated helper T cell, the B cell becomes activated as well. It proliferates and starts refining its receptors for even better binding of the antigen. Eventually, the newly produced B cells secrete the receptors into the intercellular space. The secreted receptors, called antibodies, bind the antigens, marking them for destruction

helper T cell (see below) activates the B cell. The latter signal appears only during infection. The requirement of this signal serves as one of the mechanisms of avoiding the immune response to self-antigens, own healthy cells, and large molecules of the organism. The activated B cell starts its proliferation and secretion of the BCRs into intercellular space. Secreted BCRs are called antibodies. The secreted antibodies do not have, usually, very high affinity to the pathogen. To achieve a very high affinity and specificity of the antigen binding, the activated B cells initiate the process of structural changes in the antigen-binding sites of the antibodies. It results in a great increase in the specificity and affinity of the antibodies to the pathogen in some of the activated B cells. At the final stage of their development, these B cells become large *plasma cells*. Each plasma cell secretes into the bloodstream about 2000 antibodies per second with the same antigen-binding sites (Fig. 10.1).

Part of the activated B cells is transformed into the long-living memory cells that provide a quick immune response when the same microbe or virus attacks the organism later in its life span. Each memory cell can produce only one type of antibody specific to a particular antigen.

10.1.2   Antibodies

In the absence of viral or microbial invasion, each B cell has about 10^5 identical BCRs on its surface, and there are billions of different clones of B cells with unique antigen-binding receptors. After activation, a B cell starts secreting antibodies. The antigen-binding sites of BCRs and antibodies secreted by the activated B cell are identical. Antibodies are also called *immunoglobulins* and abbreviated as *Ig*. Very unique mechanisms allow reaching this huge diversity of antibodies. We now consider these mechanisms in some detail.

Antibodies are proteins formed by four polypeptide chains, two identical heavy chains and two identical light chains. All antibodies have a similar structure diagramed in Fig. 10.2. The heavy and light chains have variable regions of about 110 amino acids, which form two identical antigen-binding sites. The light chains also have a constant region of about the same length as the variable region, while a constant region of the heavy chain is 3–4 times longer. The protein also has a flexible hinge region that allows varying separation between the two antigen-binding sites.

There are five classes of antibodies with different constant regions of heavy chains. These different constant regions are responsible for interaction with the membrane of B cell and various receptors in the cell interior. During their development, B cells switch from the production of one class of antibodies to another without changing the antigen-binding sites. At the end of the development, B cells are transformed into large plasma cells which secrete the most abundant class of antibodies, IgG.

A few unique mechanisms contribute to the huge diversity of the antigen-binding sites of antibodies. The genes coding light and heavy chains of antibodies are not identical in different B cells. These genes are assembled from a few types of segments during the development of B cells from their progenitors, the germ cells. The germ cells have many different versions of each type of segment which constitute the antibody gene in B cells (Fig. 10.3). A single copy of each segment is selected randomly to form a unique combination of these genes. This DNA rearrangement

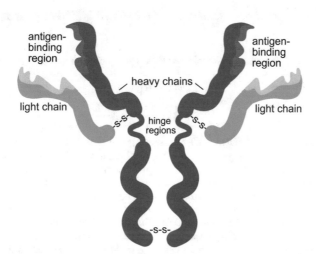

Fig. 10.2 A diagram of an antibody. The protein consists of two identical pairs of heavy and light chains (shown in dark and light blue, correspondingly). It has two identical antigen-binding sites, each formed by one light and one heavy chain. The chains are connected by disulfide bonds (shown as -S-S-), and two halves of the antibody are also held together by hydrophobic interactions between them. The hypervariable regions of the antibody, responsible for the binding of the antigens, are shown in red and pink

during B cell development is performed by site-specific recombinase V(D)J. The enzyme excises one DNA segment when forming the gene of the light chain and two segments for the gene of the heavy chain.

It should be emphasized that the formation of unique genes coding light and heavy chains in various B cells is DNA editing, a phenomenon that is unique to lymphocytes. It is very different from RNA editing, which is common in living organisms. Combinatorial assembly of the antibody gene from the V, D, and J segments during the early stage of B cell development is irreversible. This assembly creates a clone of B cells. RNA splicing, on the other hand, is used to create an mRNA segment that codes the required class of C-region of the heavy chains (see Fig. 10.3). The pattern of the splicing is changing during the later stage of development. Contrarily to this, the variable sections of light and heavy chains are randomly formed in the development, and the choice of segments is irreversible. The combinatorial assembling of both light and heavy chains creates more than a million different antigen-binding sites.

The second mechanism of antibody diversification arises from some variations in the joining of ends of DNA segments excised by V(D)J recombinase. Usually, site-specific recombinases cut and rejoin DNA segments at uniquely defined positions of specific recognition sites (see Sect. 4.3.2). It is not the case for V(D)J recombinase, which can delete or insert a few base pairs at the joint. Of course, it can result in a shift in the reading frame, and no functional polypeptide chains will be produced. If

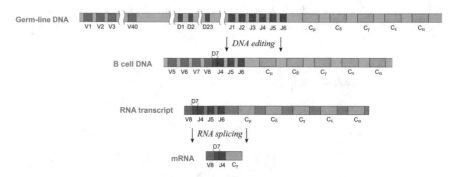

Fig. 10.3 The development of DNA and RNA segments that code the heavy chain of the antibody. The shown area of DNA of the germ-line cell (the first line) has 40 different versions of segment V, 23 versions of segment D, and 6 versions of segment J. Site-specific recombinase V(D)J cuts two DNA regions from the area. Thus, three randomly chosen V, D, and J segments form a continual DNA section that codes the variable part of the heavy chain, V8D7J4 in the example (the second line). All five versions of the C segment that code five different classes of the constant part of the heavy chain remain in the rearranged DNA. They are also present in the RNA transcript (the third line). The mRNA of the heavy chain (the fourth line) is obtained by splicing the transcript. The splicing removes introns and leaves a single copy of the C region needed for the B cell at the moment. Note that the DNA of the B cell obtains a unique combination of V, D, and J segments that specify the variable part of a heavy chain. This variable part can be combined with any of the five classes of the chain constant region by different RNA editing. The figure shows a simplified version of the process; the actual rearrangement in the DNA and RNA segments is even more complicated

this happens, the B cell can go through the second round of V(D)J recombination or die. This mechanism of diversification increases the number of different antigen-binding sites by a few orders of magnitude. Taken together, the two mechanisms listed above can result in different binding sites, whose number exceeds the number of B cells in a human body.

Still, in the random pool obtained in this way, there are usually no BCRs with very high affinity to a particular antigen. When a B cell starts proliferating, activated by the antigen and helper T cells, it accumulates point mutations in DNA segments coding the variable regions of both light and heavy chains. The frequency of these mutations is approximately a million times larger than the mutation frequency in other genes of somatic cells. Approximately one mutation occurs at each division of a B cell (B cells are capable of dividing). Randomly, antibodies with a higher affinity to the antigen appear, causing faster activation and proliferation of the B cells which produce them. The process, called receptor editing, eventually results in the production of antibodies with extremely high affinity to the antigen.

The huge variety of antibodies can protect from the invasion of nearly any microbes or viruses in the extracellular space. Antibodies do not penetrate, however, into other cells of the organism. Therefore, pathogens that are hidden inside the organism's cells are protected from antibodies. Another problem is that B cells themselves cannot prevent from being activated by self-antigens and starting self-destruction. Therefore, an additional mechanism is needed here. These two problems are solved by T cells. T cells, in turn, participate in complex interactions with dendritic cells, B cells, and macrophages. Below we very briefly outline the most important of these interactions.

10.1.3 T Cells

T cells are another major component of the adaptive immune system. There are two classes of T cells, *cytotoxic T cells* and *helper T cells*. The cells of both classes are antigen-specific, so each T cell can interact only with a particular antigen. Cytotoxic T cells directly kill cells infected by viruses or microbes. B cells cannot act on pathogens hidden inside other cells, so this function of the T cells is critically important. Helper T cells activate B cells and infected macrophages. The latter ones, when activated, kill the invading pathogens. Many features of both classes of T cells are very similar at the level of detailing in this book, so the great part of the following description refers to both cytotoxic and helper T cells.

Each T cell has many identical antigen-specific receptors on its surface, which are similar to the BCRs of B cells. The receptors have structural similarities with BCRs and similar mechanisms of diversification. There is a large pool of DNA segments coding the variable part of receptors, which are combined with the constant segments by the same V(D)J recombinase. However, unlike genes of antibodies, the genes of T cell receptors do not have a high frequency of mutations. Overall, their specificity and affinity for antigen binding cannot reach the level typical for mature

B cells. The receptors of T cells do not bind the entire molecules of an invader. Instead, they bind only oligopeptides produced by the enzymatic digestion of the pathogen's proteins.

T cells that can bind the organism's own oligopeptides have to be eliminated during their development. It occurs at the early stage of T cell development in the thymus, where the great majority of self-oligopeptides are presented. T cells that bind these peptides are inactivated or die. However, there is a small but not negligible probability that a self-oligopeptide was not presented to a particular T cell during this stage of its development. In this case, another mechanism, described below, prevents the proliferation of this cell.

T cells have to be activated before they will be able to act in the immune response. The activation of both cytotoxic and helper T cells requires their direct contact with the antigen-presenting cells, usually dendritic cells. Only after this activation do T cells proliferate and differentiate into active cytotoxic or helper cells. The activation is a very important process that, in particular, prevents the adaptive immune system from attacking the organism's own cells.

Dendritic cells, which represent a part of the innate immune system, are the key players in the activation process. The cells have many different receptors on their surface that can bind, engulf, and digest various microbes and viruses. By digesting the pathogen, dendritic cells make oligopeptides of the foreign proteins. Special proteins called the *major histocompatibility complexes* (MHC) bind these foreign oligopeptides and transfer them to the surface of dendritic cells. MHC proteins have transmembrane and extracellular regions, similar to BCRs and T cell receptors. Their extracellular region binds the oligopeptides and presents them to T cells. MHC proteins do not have specificity to the sequence of the bound oligopeptides, so they bind both the organism's own and foreign oligopeptides. The own oligopeptides are products of normal proteolytic digestion of proteins inside the cells. As we will see below, MHC proteins are also present on the surfaces of B cells and macrophages which are activated by the activated helper T cells. However, the own oligopeptides presented by MHC proteins do not bind to the T cell receptors and therefore do not activate T cells. Indeed, T cells which could bind the organism's own oligopeptides are eliminated at the earlier stage of their development. If a dendritic cell presents a foreign oligopeptide, there is a good chance of finding T cells with receptors specific to this oligopeptide. Binding an oligopeptide by the receptor gives the first signal needed for T cell activation. However, the activation also requires the second signal, which also has to come from the same dendritic cell (Fig. 10.4). This signal is provided by co-stimulatory proteins located on the surface of the dendritic cell. These proteins appear on the surface only if the dendritic cell is infected by a pathogen. The proteins interact with co-receptor proteins on the surface of the T cell. If a T cell receives signal 1 without signal 2, it is deactivated or dies. This requirement of the second signal provides an additional mechanism against the activation of T cells by self-oligopeptides.

Activated helper T cells proliferate and, consequently, can activate B cells and macrophages, their target cells (Fig. 10.5). To perform these tasks, the T cells interact with the target cells by mechanisms that are very similar to the ones that activate

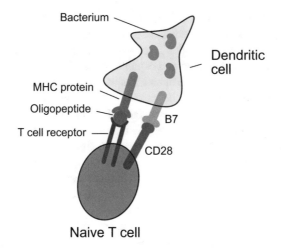

Fig. 10.4 Activation of the T cell. A naive T cell needs two signals to be activated. Usually, both signals come from an infected dendritic cell that presents oligopeptides of proteins digested in this cell. The oligopeptides have to be presented by MHC proteins on the surface of the dendritic cell. Its interaction with the T cell receptor, specific to one of the oligopeptides, provides the first signal. The second signal is provided by co-stimulatory protein B7, which appears on the surface of the dendritic cell only when it has a pathogen inside. This signal is transmitted to the T cell through interaction with the co-receptor protein CD28 located on the surface of the T cell. If both signals are received, the naive T cell develops into an activated cytotoxic or helper T cell

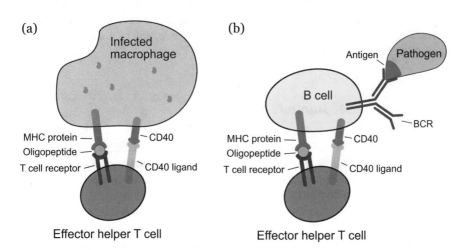

Fig. 10.5 Activation of infected macrophages and B cells by activated helper T cells. (**a**) Activation of a macrophage allows it to kill the ingested pathogens. The activation requires two signals. The first signal appears when the MHC protein presents a pathogen's oligopeptide which binds to the receptor of this oligopeptide located on the surface of the activated helper T cell. The second signal comes from the *CD40 protein*, located on the macrophage surface, when it interacts with the *CD40 ligand* located on the surface of the activated helper T cell. The CD40 protein only appears on the microphage surface when the cell is infected, while the CD40 ligand is exposed only on the surface of activated helper T cells. (**b**) Activation of a B cell requires the same two signals from an activated helper T cell, in addition to the signal caused by binding the pathogen by BCRs on the B cell surface

them. To be activated, the target cells have to present an MHC-bound foreign oligo-peptide, transferred from the interior, on their surface. Thus, the activation process is antigen-specific since only helper T cells with receptors specific to the presented foreign oligopeptide are activated and capable of activating the target cells.

The activation also requires the interaction of another co-stimulatory protein, the CD40 ligand, with its partner on the surface of the target cell, CD40 protein. The co-stimulatory protein appears on the surface of the helper T cell only if it has been properly activated. The requirement of this interaction reduces the chances of false activation of B cells and macrophages. These two signals are sufficient for the activation of macrophages (Fig. 10.5a). However, in the case of the B cell, the interactions with the helper T cell provide only two of three signals needed for its activation (Fig. 10.5b). The first signal for the B cell activation comes from the direct binding of the antigens by the BCRs on its surface.

Activated cytotoxic T cells kill the organism's cells infected by the pathogen. They do it when their receptors bind to the pathogen's oligopeptide presented by MHC proteins on the surface of the target cell. When the binding is established, the T cells use their secretory apparatus to initiate a chain of events that results in the apoptosis of the infected cell.

Similar to B cells, a fraction of activated T cells become long-living memory cells. These cells provide a fast response of T cells on repeated infection by the same pathogen. While memory B cells have very high specificity to each strain of the pathogen, T cells are not so selective. They can respond, for example, to a virus that has mutated after the first infection. Indeed, some oligopeptides, obtained from the proteins of the mutated virus, remain unchanged and can activate the memory T cells. This ability to provide a fast response to the infection of a slightly changed pathogen is an important property of the memory T cells.

10.2 The Innate Immune System

The adaptive immune system described above is very powerful, but the development of its full-scale pathogen-specific response to a new infection takes a few days. During this time, the pathogens could proliferate dramatically if the spread of the infection is not stopped or at least inhibited by the innate immune system. The innate system has very low specificity to various pathogens and is less efficient than the adaptive system, but its activation is very fast. Thus, it is needed to provide the first response to a pathogen invasion. The innate system is also important in the activation of the adaptive system, which provides a much stronger response to a specific infection. The requirement of this activation is very important as an additional control against the autoimmune response of the adaptive system. The innate system includes a few different components. Only some of them are briefly considered below.

The first and the most difficult task for the innate system is to identify pathogens. It turns out that there are common features of large groups of pathogens that can serve as markers for the innate system. Specific receptor proteins on the surface of cells of the innate system, called phagocytic cells, are capable of identifying the great majority of pathogens. But the first line of defense is provided by the skin of vertebrates.

10.2.1   Epithelium and Defensins

Skin and other epithelial surfaces provide mechanical protection against pathogenic invasion. The interior of the surfaces is covered by mucus that, in particular, prevents the adhesion of pathogens to the epithelium. The mucus also contains special small proteins called defensins. The defensins act as antimicrobial agents by killing or inactivating bacteria, multicellular parasites, and even many viruses. Some defensins have a well-pronounced 3D structure, while others are unstructured (Fig. 10.6). The positively charged defensins interact with the negatively charged bacterial membrane of the invading pathogens but not with the membrane of the host cells which is only slightly charged. The defensins also have some specificity to the type of pathogens, and the production of a particular defensin can be activated by the invaded pathogen. A variety of different mechanisms are used by defensins to kill pathogens, including making pores in their cell membranes, inhibiting DNA, RNA, and protein synthesis. The defensins are also located, in a very large amount, in the special cells of the innate immune system, *neutrophils*, which kill pathogens. Defensins were found in all vertebrates.

Fig. 10.6 Structure of human beta-defensin-2 (Sawai et al., *Biochemistry*, 2001. 40, 3810–6; PDB ID: 1FQQ). The protein consists of 41 amino acids. It has a pronounced 3D structure, as well as flexible regions. The image from RCSB PDB was obtained with the Molecular Biology Toolkit (Moreland et al., 2005, BMC Bioinformatics, 6:21)

10.2.2   Phagocytic Cells

The major players of the innate immune system of vertebrates are phagocytic cells, which recognize, engulf, and kill pathogens. There are two families of such cells, macrophages and neutrophils. Macrophages are long-living cells that have been found in all tissues of organisms. Neutrophils present in large amounts in the blood-stream and can be rapidly relocated to the site of infection. They have a relatively short lifetime. When they fight pathogens, both families of phagocytic cells act similarly.

   The phagocytic cells have many receptors on their surfaces. These receptors can bind various pathogens using their common molecular features. Large groups of bacteria have such features on their cell surface, like peptidoglycans of the cell wall. The phagocytic cells also recognize pathogens bound to antibodies. Binding the pathogens activate the signaling cascade which produces a signal for the pathogen engulfing by the membrane of the phagocytic cell (Fig. 10.7). The engulfing results in the creation of the *phagosome*, a large intracellular particle with the pathogen surrounded by the plasma membrane. When the pathogen is inside the phagosome, the phagocytic cell destroys it by various highly reactive agents injected into the phagosome. If a pathogen is too large to be engulfed by a phagocytic cell, like a large parasite, it can be surrounded by many such cells and eventually be destroyed as well. Activated phagocytic cells also produce many signaling proteins which turn on other components of the immune system.

10.2.3   Antiviral Defense

While bacteria have some specific features on their surfaces that allow their recogni-tion by the receptors of phagocytic cells, the surfaces of viruses, in general, do not have specific determinants. Therefore, dsRNA, which appears in the life cycle of many viruses (see Chap. 11) and is not a normal object in vertebrate cells, represents the major target of the innate immune system. These RNA molecules turn on the system of RNA interference (see above). The special enzymes of the system frag-ment dsRNA into segments of about 22 bp in length. The other enzymes separate

Fig. 10.7 Engulfing a pathogen by macrophage and formation of the phagosome. The pathogen will be destroyed in the phagosome

the strands of the duplexes to obtain siRNA. siRNA guides special nucleases to ssRNA molecules, whose sequence is complementary to siRNA, and special nucleases digest these ssRNA molecules. In this way, RNA interference destroys virus mRNA molecules or even its genomic RNA. In addition to this, dsRNA molecules initiate the production of special signaling molecules, interferons, by the infected cells. The interferons turn on another system of nucleases that degrades all ssRNA in the cells and inhibit the entire protein synthesis. This, in particular, halts the reproduction of viruses. In these ways, the innate immune system slows down and sometimes even stops the viral infection as well. When the virus is eliminated from the infected cell, protein synthesis is restored.

10.2.4   Natural Killer Cells

The cells infected by pathogens have to be killed to prevent the infection from spreading. Their apoptosis is initiated by the cytotoxic T cells which recognize foreign oligopeptides presented by MHC proteins on the surface of infected cells (see Sect. 10.1.3). However, some pathogens are capable of strongly diminishing the amount of MHC proteins on the surface of infected cells. Therefore, such infected cells are not recognized by the cytotoxic T cells. However, a very low level of HMC proteins on a cell surface serves as a marker for other cells of the innate system, the *natural killer cells*. The natural killers bind cells with a low level of MHC on their surface and send them the apoptotic signal, killing the cells and pathogens which invaded them.

10.2.5   Dendritic Cells

Dendritic cells are capable of recognizing a great variety of pathogens by numerous different receptors on their surface. Eventually, they engulf and digest the pathogens, similar to the phagocytic cells. But even more important is that they activate T cells, initiating the response of the adaptive immune system. This activation process was described in Sect. 10.1.3.

Chapter 11
Pathogens

11.1 Viruses

11.1.1 General Properties

Viruses are amazing creatures that occupy a separate niche between inanimate nature and living organisms since they are not able to reproduce themselves. Still, their role in life is enormous. They are a major cause of diseases of all kinds of organisms, including humans.

Viruses consist of a nucleic acid that carries their genetic information, a protein shell, and, sometimes, a few enzymes and a lipid envelope enclosing the shell. Since they are not capable of reproducing themselves, they have to use the enzymatic machinery of the invaded cell for this goal (Fig. 11.1). The genetic information of viruses is kept in either DNA or RNA molecules, which can be double-stranded or single-stranded. Viruses that cause human diseases carry double-stranded DNA (dsDNA) or single-stranded RNA (ssRNA) in their shell. Viral genomes are very compact and may consist just of a few thousand of nucleotides, although the largest viruses carry hundreds of thousands of bases. A shorter genome gives them an evolutionary advantage since it allows producing more molecules of viral DNA or RNA. Some viruses even develop overlapping genes to reduce the size of their genomes. This is achieved by shifting the reading frame in the overlapping segments of the genes, so the same DNA segment codes segments of proteins with completely different sequences of amino acids.

Viral genome codes proteins that assist its replication and are needed for the assembly of the viral shell. The genome may also have genes of proteins that help pack the viral genetic material into the shell and proteins that enhance the virus production by the cell machinery. Depending on the virus' genetic material, replication of its genome includes different steps. Replication of viral dsDNA is similar to the replication of cellular DNA. In eukaryotes, the viral DNA penetrates the nucleus

A. Vologodskii, *The Basics of Molecular Biology*,
https://doi.org/10.1007/978-3-031-19404-7_11

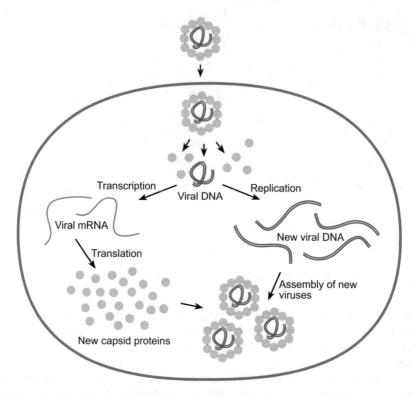

Fig. 11.1 A simplified picture of the virus life cycle. The double-stranded DNA of this hypotheti-cal virus is coated by the protein shell. The shell is disassembled after the virus penetrates the cell. The viral DNA is transcribed to form mRNA and is replicated to form DNA for new viruses. Translation of the viral mRNA produces new shell proteins needed for the self-assembly of new viruses, which eventually leave the cell. Thousands of new viruses can be produced in a single cell infected by a single virus particle

where it is transcribed and replicated. Then the viral mRNA molecules leave the nuclei and serve for the production of the viral proteins.

To penetrate inside a cell, the virus first has to bind to its outer membrane. This binding is mediated by specific proteins on the virus shell (or envelope) which have a high affinity to certain receptors on the cell's membrane. Since the receptors are specific for the cell type, the virus can penetrate only into a particular type of cell. For example, the virus of hepatitis B can only attack certain cells of the liver called *hepatocytes*. There are various mechanisms by which viruses can disrupt the cell membrane and penetrate the cytoplasm.

Very high productivity and short genome allow viruses to have error-prone rep-lication which results in a very high mutation rate, up to 1 mutation per 10,000 nucleotides per replication cycle. For RNA viruses, such a very high mutation rate is due to virus-coded *reverse transcriptase* (see below) which synthesizes DNA on an RNA template with very low accuracy. Under such a high mutation rate, a large fraction of new viral genomes receive deleterious mutations, but it is not a problem for the virus due to a huge number of new particles produced in a single cell from a single virus particle. On the other hand, such a mutation rate gives viruses a great evolutionary advantage, since their proteins constantly change. This mutation rate enormously complicates the antiviral defense. Indeed, the development of a full-scale response of the adaptive immune system takes a few days, and over this time, the virus may be able to change its antigen determinants. So, the immune system has to develop new antibodies for the modified virus, and so on.

11.1.2 RNA Viruses

Viruses that carry ssRNA as genetic material have first of all to convert their RNA to dsDNA. Since eukaryotic cells do not have, normally, enzymes capable of synthesizing DNA on an RNA template, the viruses have to code the corresponding enzymes in their genome. Some of these viruses carry the enzyme, the reverse transcriptase, in their shells, while others use the cell machinery and their RNA to synthesize the enzymes after penetration into the cell. First, the enzyme makes the RNA-DNA hybrid and then uses it as a template to synthesize the corresponding dsDNA. The obtained DNA molecules serve for the production of viral mRNA and RNA molecules for new viral particles.

Among viruses with single-stranded RNA, there is a family of retroviruses, whose life cycle includes the integration of the viral DNA into the host genome. A few retroviruses infect humans, and HIV is one of them. The viral DNA moves into the nucleus where another viral enzyme, *integrase*, inserts it into the DNA of the host cell, similar to the insertion of transposons. Thus, the viral DNA becomes a part of the host genome. It is transcribed to obtain mRNA molecules that are used for the

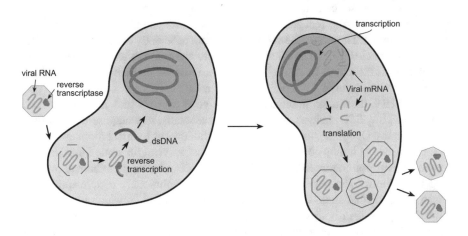

Fig. 11.2 The life cycle of retroviruses. After entering the cell, the virus destroys its shell to release its genetic material, ssRNA. The virus-coded reverse transcriptase synthesizes the RNA/DNA hybrid and then dsDNA. The obtained DNA and the viral enzyme integrase enter the nucleus of the host cell where the enzyme inserts the viral DNA into the host genome. The viral segment is intensively transcribed to obtain mRNA and full viral RNA for the new viruses. mRNA is used for the production of new viral proteins. Self-assembled viral particles leave the cell

production of viral proteins. The transcription also gives the full viral RNA needed for new viruses (Fig. 11.2). The reverse transcriptase is an extremely error-prone polymerase and, therefore, introduces a lot of mistakes during the replication. This results in a very high mutation rate of the genomes of RNA viruses.

Usually, retroviral DNA is incorporated into the genome of a certain type of somatic cells occupied by the virus. These cells have to have specific receptors on their surface that are used for the attachment of the virus. For example, HIV can only enter certain types of T cells and macrophages, which carry the receptor CD4 on their surface. The receptor has a very high affinity to viral protein gp120 exposed on the virus surface. Without binding to CD4, the virus cannot enter a cell, so only specific types of cells can have a viral DNA inserted into their genome. This gives hope to finding a cure for the virus which would eradicate it from infected individuals since the lifetime of the infected cells is limited.

11.1.3   Evolution of Viruses

A very high rate of mutations changes the pathogenicity of viruses. Over the last 100 years, a few such events had very bad consequences. The worst case happened in 1918 due to a new strain of influenza virus which caused the death of at least 20 million people worldwide (although bad living conditions and malnutrition, brought by World War I, made a very large contribution to this mortality). Normally, the influenza virus causes relatively mild flu in adults, and its mortality, on average, is about 0.1% in people with confirmed influenza infection. In some groups of people, particularly those over 65 years old, the mortality can be a few times higher, however.

In general, viruses can infect only one type of species, because even in evolutionarily close species, the receptors on the surface of functionally identical cells are slightly different. Therefore, the same virus cannot efficiently bind the cell receptors of different species. Even a small difference between human's and monkey's receptors can be sufficient to be distinguished by a virus, as is the case for HIV. Still, sometimes viruses jump from one species to another. There is a virus that is very similar to HIV, the simian immunodeficiency virus (SIV), which infects monkeys. Similar to HIV, SIV changes very fast and, eventually, can acquire the ability to infect humans. Of course, the ability of the modified virus to infect monkeys will be compromised but it may be sufficient for the virus' survival for some time. Data are supporting the notion that SIV jumped a few times from monkey to human over the twentieth century. Jumps of viruses between more evolutionary remote species are much less probable since such jumps require a form of virus which is capable of reproducing itself in both species. Still, it is generally accepted that the *influenza virus* and *coronavirus* have jumped from animals to humans a few times over the last decades, causing cold and respiratory diseases in many countries.

In 2019, a new strain of coronavirus caused a worldwide pandemic. Usually, coronavirus causes only a mild respiratory disease or common colds in adults. In general, the disease is less damaging than the one caused by an influenza virus. However, from time to time, strains of coronavirus appear which cause more severe

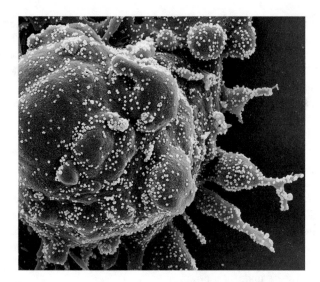

Fig. 11.3 Electron micrograph of a cell (green) heavily infected with particles of the COVID-19 virus (yellow), isolated from a patient sample. The size of the virus particle is close to 100 nm. The micrograph is colored by a computer. (Courtesy: National Institute of Allergy and Infectious Diseases)

consequences. The strain of 2019, called COVID-19 (Fig. 11.3), caused higher-than-average mortality. The disease spreads out faster than a usual common cold, and some important questions about it remain to be answered. The new virus binds the cell surface receptor called ACE2 which is abundant in lung cells, some cells of the gastrointestinal organs, and the kidneys. It can also cause damage to the cardio-vascular system. Still, the disease is asymptomatic or nearly asymptomatic for the majority of infected people.

The fast spread of the COVID-19 virus around the world created an enormous problem. The efforts of many scientists are directed now to the search for medications or making a vaccine against the virus. In the first half of 2020, a few vaccines were created and vaccination started in many countries. This pandemic clearly showed that such usually not-so-dangerous viruses require much more attention.

11.1.4 Prevention and Cure

Since the replication of viruses mainly uses the replication machinery of the infected cells, it is difficult to find antiviral drugs that do not harm the infected organism. Another factor that makes the search for antiviral drugs so difficult is the fast change of viral genomes and, correspondingly, the viral proteins.

Under such conditions, vaccination so far was the most effective way to fight diseases caused by viruses. It was very successful against very dangerous human viruses such as smallpox and poliovirus. Both of these viruses have been practically eradicated due to global vaccination. Broad vaccination of girls and boys against a few strains of human papillomavirus, which can cause cervical cancer, is underway now.

Vaccination is not so effective against some other viruses. The main reason for failures to make an efficient vaccine is the fast changes in the viruses. The influenza virus can serve as a good example. Although new vaccines against the virus are designed and distributed twice per year in the USA, their effect is limited. Although a few vaccines against COVID-19 are used now, it remains to be seen for how long their effect extends. Despite many years of efforts, all attempts to create a vaccine against HIV have been unsuccessful.

Meanwhile, remarkable results were achieved in the search for a cure for HIV. Researchers were able to find drugs that inhibit each of the viral enzymes. If an infected person takes a mixture of these drugs regularly, they nearly completely suppress the disease caused by the virus. Still, the virus remains in the infected cells in the latent form, since its genome is inserted in the genome of these cells. To completely eradicate the virus, these cells have to be eliminated. It may be possible, since their lifetime is limited, and they are eventually recycled by apoptosis. The eliminated cells could be replaced by new T cells and macrophages produced from uninfected specialized stem cells (see Fig. 8.14). The efforts to develop the corresponding therapy are underway now.

11.2   Bacteria

About a 100 various bacteria belong to human pathogens, and they cause various diseases in many organs. Pathogenic bacteria vary greatly in their features. Some of them enter specific cells of the human body and can grow and replicate only in such cells. Others live as extracellular pathogens. Some bacteria can only infect humans, while others are capable of living in a wide variety of organisms. There is even a bacterium, *P. aeruginosa*, which infects both animals and plants. In humans it causes sepsis.

11.2.1   Invading into a Human Body

Bacteria use a variety of ways to enter the human body. Epithelial cells do not protect the internal organs as well as the skin, and many bacteria enter through these cells. This is the case, in particular, for many bacteria that enter through the respiratory tract, like *M. tuberculosis*, *S. pneumoniae*, and *H. influenzae*, and through virginal epithelia, like *N. gonorrhoeae* and *C. trachomatis*. Others can enter through skin cuts, like *Staphylococcus aureus* (which causes sepsis and some other diseases). Many bacteria penetrate through the digestive tract with food or water, like *Vibrio cholerae*, which causes cholera; *Salmonella*, a common cause of food poisoning; and pathogenic *E. coli*. Some bacteria, like *B. burgdorferi* which causes Lyme disease and *Y. pestis* which causes bubonic plague, use blood-feeding ticks that infect animals. The bacteria which live in the tick penetrate the animal and infect it during a long feeding session.

After entering the extracellular interior of the body, many bacteria need to penetrate specific cells. To do this, they first bind to certain receptors on the cell surface. Surface receptors are usually specific to a cell type, so the presence of the needed receptors on the cell surface directs the invading bacteria to these specific cells. After binding to the receptors, the bacterium needs to penetrate the cytoplasm. It usually occurs by *phagocytosis*, a process of engulfing an invading bacterium or virus by the membrane of the host cell to form the phagosome (see Fig. 10.7). Normally, the phagosome facilitates the destruction of the invader by injecting aggressive digesting enzymes into it. There are special phagocytic cells of the immune system which engulf the pathogen, create phagosome around it, and digest it (see Sect. 10.2.2). Some bacteria, however, enter these cells by phagocytosis and manage to avoid distraction. They use the phagosome as a protective shell and create inside it the needed microenvironment. Other bacteria invade nonphagocytic cells. To force phagocytosis in such cases, they secrete special proteins that bind to the cell surface and initiate the process.

11.2.2 Pathogenesis

In one way or another, pathogenic bacteria change the behavior of the host in a way that is beneficial to the pathogen. For this goal, they use toxic proteins, *toxins*. The majority of toxins affect signaling cascades of the host cells, which results in an improper concentration of some intracellular substances critical for cell life. For example, a few bacteria, including *Vibrio cholerae*, produce a toxin that eventually increases the concentration of cyclic AMP inside the cell. Cyclic AMP is a key component of many signaling cascades (see Sect. 7.4). In particular, its overproduction causes the release of ions and water into the intestine interior, resulting in watery diarrhea. Diarrhea facilitates the spread of the bacteria to other hosts through water contamination by the bacteria.

Some extracellular pathogenic bacteria secrete toxic proteins into the extracellular space. The proteins can compromise the immune response of the infected organism. They can also facilitate the spread of the infection. An example of this is the colonization of the respiratory tract by *B. pertussis* which causes coughing. Other extracellular bacteria inject toxins directly into the host cells through a syringe-like apparatus.

Bacterial toxins are coded by special virulence genes. The genes are usually clustered in the bacterial genome but can be also located in a virus that infects the bacterium (bacteriophage), in a transposon integrated into the bacterial DNA, or in a bacterial plasmid. Clustering of the virulence genes greatly facilitates their transfer from one bacterium to another. This transfer occurs through the horizontal gene transfer (see Sect. 4.3.2) if the genes are located in the bacterial genome. The horizontal gene transfer occurs quite often in bacteria, and due to such transfer, different strains of the same bacteria can exist without or with the virulence genes. An example of this is the bacterium *E. coli*, which can be nonpathogenic and even useful for the digestion of food in our stomach if it does not have virulence genes. However, the strain with the virulence genes causes serious food poisoning in humans. The horizontal transfer of virulence genes can occur between different species of bacteria as well.

Sometimes a bacterial infection does not cause immediate harm to the organism, and bacteria can live inside the human body for many years. This is the case for *H. pylori*, the bacterium which colonizes the digestive tract. *H. pylori* was discovered only at the end of the last century, although it is widely spread in the human population around the world. The infection by the bacterium remains asymptomatic for years and even decades before it results in heavy consequences, first of all, stomach ulcers. Ulcers are a serious disease by itself and can eventually cause stomach cancer. There are unconfirmed data that the bacteria are associated with many other human diseases as well.

11.2.3   Antibiotics

Bacteria are well separated from vertebrates evolutionarily. Due to this separation, bacterial enzymes, which perform replication, transcription, and translation, are somehow different from those in the infected hosts. This allows for designing antibacterial drugs that are not so harmful to humans and animals. So far, however, the most potent antibacterial drugs, *antibiotics*, were created by nature. Antibiotics are natural substances, developed, first of all, by soil bacteria as a defense mechanism against attacks by other bacteria. The discovery of the first antibiotic, penicillin, before the Second World War, transformed the fight against bacterial infections. Now more than a dozen various classes of antibiotics have been found. Currently, the term "antibiotic" is applied to all medications that kill bacteria or inhibit their growth, both produced by microorganisms and made artificially.

In the second half of the twentieth century, antibiotics brought enormous success in the fight against many diseases caused by bacterial infections. However, today many pathogenic bacteria have acquired resistance to antibiotics used in medical practice. Since natural antibiotics were used by bacteria and some other organisms for millions of years, elaborated mechanisms of antibiotic resistance have been created in the bacterial world (Fig. 11.4). All mechanisms are based on special proteins coded in a segment of the bacterial genome or a bacterial plasmid. There are three major mechanisms of resistance. The bacterium can modify the structure of the enzyme interacting with the antibiotic so that the antibiotic does not bind to the enzyme anymore. It can pump out the antibiotic by a special transporter embedded in the bacterial wall. There are both the transporters specific to a particular drug and transporters that pump out a rather broad class of antibiotics. Also, the bacterium can acquire the gene of an enzyme that digests the antibiotic. Excessive usage of antibiotics has resulted in a broad spread of antibiotic-resistance genes among both

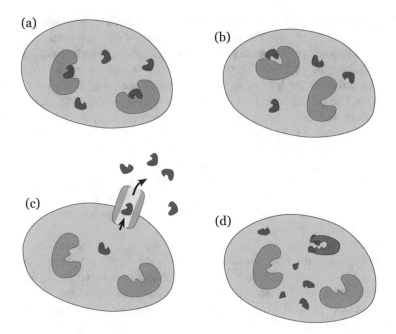

Fig. 11.4 Mechanisms of antibiotic resistance in bacteria. (**a**) Normally, the antibiotic (brown) binds to a critical enzyme of the bacterium (green) and inhibits its action. (**b**) Mutation in the enzyme changes the shape of its binding site so that the antibiotic does not bind with the enzyme and cannot inhibit it. (**c**) The bacterium acquires a transporter that pumps out the antibiotic from the cell. (**d**) The bacterium acquires an enzyme (red) that digests the antibiotic

nonpathogenic and pathogenic bacteria living in humans and domestic animals. This spread occurs mainly through horizontal gene transfer. Therefore, antibiotics become less and less effective against pathogenic bacteria. Unfortunately, very few new antibiotics were found over recent years, and new ways to fight bacterial infections are desperately needed.

11.3 Eukaryotic Parasites

Eukaryotic pathogens range from unicellular fungi to large, complex multicellular organisms, like parasitic worms. In the context of this book, the unicellular parasites are the most important, and below we will consider only these species. The number of unicellular parasites causing diseases in humans is much smaller than the number of pathogenic bacteria, but they are widely spread in humans, especially in countries with tropical climates.

Some of the unicellular eukaryotic parasites are extracellular, while others are intracellular. They have a complex organization that helps them to survive inside their hosts. A good example of this complexity presents *Plasmodium*, which causes malaria, one of the most devastating diseases. Around half of a million people die from the disease every year. The parasite exists in many different forms. It is spread through a certain type of infected female mosquitoes that bite humans and infect them while soaking the blood. Through the blood vessels, the parasites travel to the host liver and enter the liver cells. They reproduce asexually in the cells to form *merozoites*. The merozoites do not represent a completely matured form of the parasite, although they are capable of asexual reproduction. They multiply in the liver cells and then infect red blood cells where they go through a few more cycles of such reproduction. In the blood cells, some merozoites develop into precursors of male and female gametes. These precursors can mature only when they again occur in the mosquito after the insect bites the human host again. The gamete precursors develop there and fuse to form zygotes, which grow up to complete the reproduction cycle. Malaria can cause multiple complications in the respiratory tract, kidney failure, and damage to other organs.

Like all other pathogens, eukaryotic parasites have to delude the adaptive immune system of the host, which tries to kill invaders. If viruses use a very high rate of mutations in their genome to complicate the immune response, eukaryotic parasites develop special, sometimes very sophisticated, mechanisms to protect themselves from the host's immune system. An impressive example of such a mechanism gives the extracellular parasite *T. brucei*, which causes African sleeping disease. The parasite is covered by a protein that serves as an antigen for the host's immune system. In a few days, the immune system develops antibodies to this antigen and nearly eliminates the infection. The parasite has, however, about 1000 different genes for the surface protein. Just before the complete elimination of the parasite by the body, *T. brucei* switches to the production of another version of the surface protein which is not recognized by the same antibodies. Therefore, the number of parasites in the body increases again, and so on. This is why the disease caused by the parasite has a cyclic character. Now drugs have been developed against the disease caused by *T. crucei*. If not treated, the disease is fatal.

It is more difficult to find drugs against eukaryotic parasites than against pathogenic bacteria since many of their enzymes are closer to the corresponding human ones. Although some drugs help to fight the diseases caused by eukaryotic parasites, success is still limited. There are long-lasting efforts to develop a vaccine against *Plasmodium*, causing malaria, but so far these efforts have not been successful. Practically, there is no vaccine against any of the eukaryotic parasites. A few eukaryotic parasites are transmitted through bites of insects, like mosquitoes and tsetse flies. Therefore, one of the effective ways to limit the spread of infections is protection from insects carrying parasites.

Chapter 12
Cancer

12.1 General Remarks

The cells of multicellular organisms work in remarkable coordination and are ready to sacrifice their life if the organism needs it. This coordination is possible through an extremely sophisticated network of signaling between the cells. Cells receive signals from other cells which cause them to grow or rest, divide or die. However, the cells operate with individual molecules, and their behavior is disturbed by thermal motion. Due to this noise, errors during DNA replication are unavoidable. In addition to these errors, there are external factors such as UV radiation and harmful substances in the environment that damage DNA molecules. Normally, the great majority of such damages are repaired by the elaborated system of DNA repair enzymes, but a small fraction of the damages become inheritable mutations. To reduce the effects of mutations, each cell also has special systems that check that its life proceeds correctly. If deviations from the normal behavior accumulate, the internal controlling system develops signals causing cell apoptosis. Overall, the cells are well protected from harmful mutations. Still, the mutations accumulate and eventually can cause improper responses of the cell to the signals from its internal controlling system and the surrounding cells. In such cases, the cell can start proliferating rather than undergo apoptosis. More often, the system of DNA reparation is compromised in such cells, and they accumulate new mutations faster than normal cells. More and more of the internal controls stop working properly. The progenies of the cell continue to divide regardless of the internal and external signals. This colony of dividing cells is called cancer.

Cancer is one of the major causes of human death. However, it is not the only reason that this chapter is dedicated to the disease. Studying cancer researchers achieved a better understanding of many processes in biology, and, of course, this understanding helps to fight the disease. It is well established today that the reason for cancer is the accumulation of DNA mutations, which change cell behavior. In some cases, rare in humans, pathogens can bring the mutated copies of human genes

A. Vologodskii, *The Basics of Molecular Biology*,
https://doi.org/10.1007/978-3-031-19404-7_12

into the cells and, in this way, help in the development of cancer. However, this remains to be a side effect of the infection. Cancer is a genetic disease caused by damage in the cell genome which disrupts the normal socially responsible behavior of the cell that initiates the disease.

In this chapter, we consider some general properties of the disease, its specifics at the molecular level, its prevention, and methods of treatment.

12.2   Basic Features of Cancer

There are two major features of cancer cells: (1) they divide out of control, and (2) they invade territories normally occupied by other cells. It is the second property, *malignancy*, which makes cancer incurable. Initially, all cancers are noninvasive and usually completely curable (they are also called benign tumors). Only years later, they develop the ability to invade surrounding tissues and become real cancers. Thus, early diagnostics is very important for successful cancer treatment.

The probability of acquiring cancer increases with age. At first glance, it does not look surprising since mutations accumulate with the organism's age, and, correspondingly, the probability of cancer development has to grow over time. However, if a single mutation could cause cancer, the probability of acquiring the disease in any given year would be the same. But the observations show a different picture. The probability of cancer grows fast with aging. This means that a few mutations are needed for the full-scale development of the disease.

Although only a tiny fraction of mutations can initiate uncontrolled cell division, simple estimation shows that the appearance of such events is not so surprising. About 10^{16} cell divisions occur in the human body over a lifetime. Since the probability of a mutation in a single gene is close to 10^{-6} per cell division (see Chap. 6), about 10^{10} various mutated copies of each gene appear in the organism over its lifetime. Therefore, the probability of receiving in a single cell a mutation that stimulates the uncontrolled proliferation of this cell is not negligibly small. Still, both individual cells and the organism as a whole have many defense mechanisms against uncontrolled cell proliferation. These mechanisms cause the death of the mutated cells in the great majority of cases, but sometimes a cell with cancer-related mutations survives. It starts a colony that accumulates further mutations at an increasing rate since normal controls are compromised in this colony of cells. Many cells in the colony die, but others keep proliferating and acquire further cancer-related mutations. The initial clone splits into subclones with different mutations, and these subclones compete with one another for the organism's recourses. Eventually, one or a few subclones prevail over the others and can be transformed into full-scale cancer. This transformation is a long process, however, since it requires the accumulation of about ten different cancer-related mutations (Fig. 12.1). The evolution of the initial tumor may result in the formation of a few genetically different malignant colonies, and each of them spreads through the body tissues, a process called *metastasis*. This genetic diversity makes the treatment of cancer more difficult. In some cases, epigenetic changes in the cell chromosome contribute to cancer development.

Fig. 12.1 Development of cancer. It starts from a single cell that acquired a few mutations promoting uncontrolled proliferation. The growing clone of cells can later receive more mutations, which make the cells malignant, so they become capable of leaving their initial territory and colonize other tissues

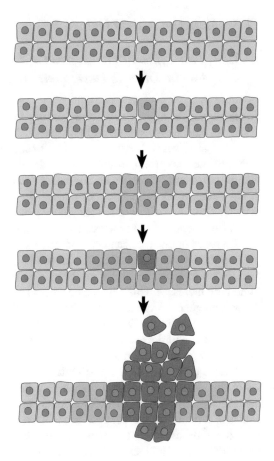

The great majority of normal human cells cannot divide indefinitely due to the shortening of the telomeres at the chromosome ends. Only cells that express special enzyme telomerase, which prevents chromosome shortening, do not have this restriction. Cancer cells have to overcome this obstacle, as well. They usually maintain the activity of the telomerase or use homology recombination to avoid chromosome shortening.

Usually, cancer causes death only when its cells give rise to metastasis. Metastasis has a few critical obstacles that cancer cells have to overcome. The cells have to leave a local tissue where they initially appeared and enter the system of blood vessels. Then they have to leave the vessels at distant parts of the body. Finally, they have to colonize new tissues. Each of these steps has to overcome defense obstacles that do not allow the spreading of normal cells. All tissues have adhesive mechanisms that keep their cells together. These mechanisms have to be destroyed in the cancer colony to initiate metastasis. This means the cancer cells have to acquire the needed mutations in their DNA. The other steps, colonizing a new tissue, are even more difficult. In particular, the invading cells occur in a very different environment where it is difficult for them to survive. Only a tiny fraction of cells, which try to do it, succeed. This is why the development of metastasis from a localized tumor usually takes years.

12.3   Cancer-Critical Genes

12.3.1   Two Types of Genes Are Critical
for the Disease Development

To understand the molecular mechanisms of cancer, one has to identify genes whose variations can be responsible for the disease. This task has been pretty much solved over the last 40 years. Genes responsible for cancer can be divided into two groups. Genes of the first group promote cancer development if their activity is increased due to mutations. These genes are called *proto-oncogenes*. Genes of the second group increase the probability of cancer development if mutations reduce their activity. Thus, in their normal form, these genes suppress the development of cancers. Correspondingly, they are called *tumor suppressor genes*.

Two types of changes can cause increased activity of a proto-oncogene. First, a point mutation can change the protein, and its new form can be much more active than the normal one. Second, a protein can be overproduced in its normal form, so its activity increases. This may happen due to a mutation in the regulatory region of the protein gene, or due to the multiplication of the gene. The latter event can happen as a mistake in the DNA replication system. Tumor suppressor genes control the progression of the cell cycle. Therefore, the loss of one of these genes damages the control system and can allow inappropriate cell division.

Mutated proto-oncogene can stimulate cancer formation even if one of two copies of the gene acquires the corresponding mutation. Indeed, it can be enough to increase the activity of the gene's product many times. Contrary to this, only if both copies of a tumor suppressor gene are inactivated, the activity of the coded protein is suppressed sufficiently to promote the formation of cancer. It is very rare that both copies of the gene will be inactivated in a single cell during the organism's lifetime. If one mutant copy of the gene was inherited, complete inactivation of the gene only requires inactivation of the second copy (Fig. 12.2). This can happen with a notable probability during the lifetime of the individual. The inactivation can also occur through epigenetic changes. In general, it is much easier to destroy something than to change its function. Correspondingly, many more mutations can inactivate a tumor suppressor gene than convert a proto-oncogene into an oncogene. Thus, individuals with inherited mutations in one copy of a tumor suppressor gene are in the group at a high risk of cancer development.

12.3.2   Changes in Regulatory Pathways

Direct sequencing of DNA from cancer cells of many individuals and comparison with DNA sequences from their healthy cells have helped to identify cancer-critical genes. About 300 such genes have been found in humans. In their normal form, these genes perform a wide spectrum of functions. Among these proteins are

(a)   (b)   (c)

Individual inherited two   Individual inherited one   Individual inherited two
normal copies of the gene   copy of mutant gene   normal copies of the gene

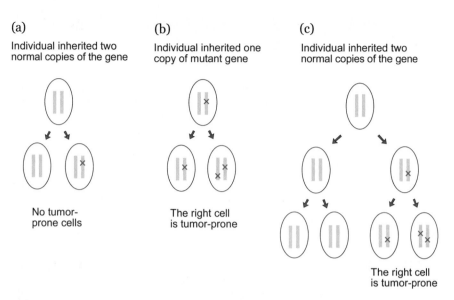

No tumor-   The right cell
prone cells   is tumor-prone

                                                            The right cell
                                                            is tumor-prone

Fig. 12.2 Inactivation of the tumor suppressor gene requires the inactivation of both copies. (**a**) Mutation in one of two normal copies of the gene does not make the cell tumor-prone. (**b**) Each cell of an individual inherited one mutated copy of the gene. A mutation in the second copy of the gene can happen with a relatively high probability. This completely inactivates the gene. (**c**) Occasional mutations or silencing of both normal copies of the gene is possible but highly improbable. These events make the cell tumor-prone

transcription regulators, signal proteins, transmembrane receptors, protein kinases, cell-cell adhesion molecules, cell-cycle controllers, apoptosis regulators, DNA repair enzymes, and others. Thus, the picture of the disease looked very complex. However, accumulating knowledge of the normal functions of these proteins helped to understand connections between the cancer-critical genes. It was found that in the great majority of tumors, three regulatory pathways were damaged. Correspondingly, cancer-critical mutations changed one or other proteins involved in each pathway. The first of those, called *Rb pathway*, initiates the cell division cycle. The second pathway, *RTK/Ras/PI3K,* transmits signals for cell growth and division from other cells. The third one, *p53 pathway*, regulates the cell responses to DNA damage and *cell stress*, the states in which one or another aspect of cell metabolism is out of normal boundaries. In nearly all tumors, each of these three fundamental controls is disrupted due to mutations in one of the many proteins involved in each pathway.

p53 protein and the pathway called after its major component are particularly important. Damages in this pathway were found in 96% of all cancers. The gene of p53 itself is mutated in about 50% of cancers, the highest fraction among all cancer-critical genes. The healthy protein stops the cell cycle when its DNA is damaged or the cell is stressed, to give it time to repair a damaged system. If the repair is not successful, p53 develops signals for cell apoptosis (Fig. 12.3).

p53 is a transcription regulator. It binds DNA as a tetramer (a complex of four proteins) and does not function properly, even if only one of the four subunits has a

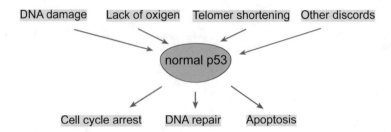

Fig. 12.3 Functions of p53 tumor suppressor. Under normal conditions, p53 has a very low concentration inside the cell. DNA damage and a broad range of abnormalities in the cell cycle sharply increase p53 concentration. The protein causes the arrest of the cell cycle to give the cell time to repair damages and recover from the stress. If this does not help, p53 initiates apoptosis of the cell

mutation. Therefore, a mutation in only one allele of the p53 gene can be sufficient for nearly full inactivation of the protein. This feature of the p53 gene is in contrast with other tumor suppressor genes. This is why people who inherit only one functional copy of the *p53* gene have a high probability of developing cancers.

Although normal functioning of the three pathways which regulate cell growth, survival, and division is critically important for cancer prevention, cells of some tissues have other pathways with similar functions. Mutations in the genes responsible for those more cell-specific pathways can be cancer-critical as well.

12.4   External Causes of Cancers

Remarkable achievements in understanding the molecular mechanisms of cancer over the last decades substantially reduced the incidences of many cancers and the total mortality from the disease. Still, the incidences of cancer cannot be prevented completely because mutations in DNA during its replication and repair are unavoidable. It is possible, however, to reduce the risk of the disease because certain external factors also contribute to its development. These factors can be divided into three categories listed below.

12.4.1   Carcinogens

Since cancer results, first of all, from mutations in DNA, the chemical substances that penetrate the body and cause DNA mutations stimulate the development of the disease. Many of these substances, called *carcinogens*, have been identified. Each carcinogen accumulates in a specific organ where it penetrates with food, through the skin, or breathing. Correspondingly, specific carcinogens promote cancers in different tissues. Tobacco smoke is the most important carcinogen, which is responsible for lung and bronchus cancer. It is difficult to obtain an accurate estimate of how much smoking increases the risk of cancer, although all data indicate that the increase is substantial. Another very strong carcinogen is aflatoxin B1 which is produced by fungi contaminating tropical peanuts. The toxin strongly increases the rate of liver cancer. Among other carcinogens that can be mentioned include benzene, arsenic, and asbestos.

The majority of carcinogens are not aggressive chemicals. They become damaging reagents only after certain modifications by specific enzymes in the liver. Normally, these enzymes help to convert toxins into harmless substances. However, they are also capable of modifying certain carcinogens, converting them into highly reactive products.

12.4.2 Lifestyle

Numerous data indicate that a healthy lifestyle notably reduces the risk of cancer. A sufficient amount of exercise and a proper diet seem to be components of such a lifestyle. Even more important is the amount of food a person consumes. Obesity is associated with a substantial increase in the incidence of many cancers. However, the data on the influence of external factors on cancer incidences should be considered with a certain caution. For example, the well-established correlation between obesity and cancer incidences does not mean that obesity is responsible for the risk of cancer. Some other factors which correlate with obesity may be the real reasons for the higher risk of cancers among overweight people.

12.4.3 Viruses and Other Pathogens

Cancer in humans is not an infectious disease; it is not transmitted by pathogens from one person to another. Still, some viruses and bacteria promote the development of certain cancers. A well-studied case of cancer of a woman's uterine cervix presents an example of the most direct effect of DNA viruses on cancer development. The papillomavirus infects the cervical epithelium and, for a long time, replicates its DNA simultaneously with the cell chromosomes, maintaining itself in a latent phase. However, to multiply its number, the virus has to force the cell machinery to replicate the viral DNA much more often, and it eventually switches to the regime where the replication of viral DNA occurs independently from the cell cycle. The switch is instrumented by viral proteins that bind and inactivate the cellular proteins controlling the cell cycle. This switch and the fast formation of virus particles are relatively harmless for the organ. But occasionally, the genes which control the involved viral proteins, E6 and E7, can be integrated into the cell chromosome and become active there. The proteins, now coded by the cell chromosome, can disrupt cell-cycle control. It was found that proteins E6 and E7 do it by binding two major tumor suppressor proteins, Rb and p53 (Fig. 12.4). Thus, we have here an

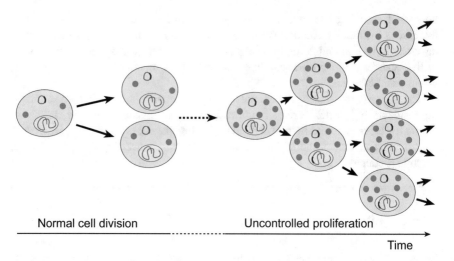

Normal cell division   Uncontrolled proliferation

Time

Fig. 12.4 The uterine cervix cancer is assisted by papillomaviruses. The virus is relatively harm-less until two of its genes, coding viral proteins E6 and E7 (shown in orange), are accidentally integrated into the host's DNA (red segment). If the acquired genes are expressed by the cell, they produce a large amount of E6 and E7 and suppress the control of the cell cycle. This promotes unrestricted cell proliferation

example of oncogenes that are brought into the cellular genome by viruses. It is worth emphasizing, however, that integration of these genes into the cellular genome is only an accidental step and is not a part of the virus life cycle. Although infection by the virus strongly increases the risk of cancer, it is insufficient for tumor development. Additional mutations in cellular DNA are also needed.

The cancer of the uterine cervix contributes to 6% of all human cancers, so it is an essential health factor. Now the risk of the disease can be reduced greatly by vaccination against papillomaviruses. Mass vaccination of girls before their age of puberty has been started in many countries.

Other viruses increase the risk of cancers in less direct ways. Among those are hepatitis B and C viruses, which promote the development of liver cancer. The viruses cause permanent inflammation of the liver, which strongly stimulates cell division. Since cell division is a major source of DNA mutations, the risk of liver cancer is increased for the infected people.

12.5   Cures of Cancers

The most common treatments for cancer consist of radiation and drugs that damage DNA in cancer cells, in addition to surgical removal of the affected tissues. Radiation and drugs damage DNA in healthy cells as well, so they are toxic to the organism. Still, they have certain specificity to the cancer cells, and despite their toxicity, they are widely used in curing the disease. The roots of this specificity were understood many years after the treatments were introduced into the medical practice. The specificity is based on the genetic instability of cancer cells. Normal cells have very sophisticated controls of DNA damage. If the damage is detected during DNA replication, it stops and special systems try to repair the damaged DNA molecule. If the repair fails, p53 turns on the apoptosis of the cell. The control systems do not work properly in cancer cells, so DNA replication in these cells usually proceeds regardless of strong DNA damage induced by the DNA-damaging treatments. Such replication increases the DNA damage to the extent incompatible with the cell life and the progeny of the cancer cells die very soon after the division of the parental cells.

Still, the mortality from the disease remains high, and there is a permanent search for better treatments. Over the last decades, this search has brought very promising results due to a remarkable improvement in understanding cancer biology. New, much more elaborated treatments for the disease are appearing. The general goal of these treatments is to kill cancer cells without making any harm to normal cells. Some examples of these approaches are shortly described below.

12.5.1 Inhibiting of Proteins Crucial for Cancer Development

Since cancers are based on mutations in cellular DNA, they are all different. However, they can be combined into groups where mutations, crucial for cancer development, affect the same proteins or protein systems. Therefore, it seems that the most direct way to kill cancer is to develop drugs that inhibit these proteins. Of course, it is important to choose a protein whose activity is critically important for the cancer cells, but the reduced activity of their unmutated precursors is tolerable for the normal cells. It was found that in many cancers, such oncogenic proteins exist. It is possible to find substances that specifically inhibit these proteins. This approach is under development for a few cancers, and the first results look very promising.

The development of the treatment for *chronic myelogenous leukemia* gives a good example of the new approach to disease treatment. This cancer results, first of all, from the mutation in a tyrosine kinase, called *Abl*, which is involved in cell signaling. In this case, the mutation represents a substitution of the C-terminal portion of Abl by the N-terminal portion of another protein. The chimeric protein maintains the function of Abl and becomes hyperactive. By phosphorylating another protein in the signaling chain, the chimeric protein stimulates the uncontrolled proliferation of the cells and prevents them from apoptosis. Thus, a drug that would block the binding of chimeric Abl with its target protein could block the signal transduction (Fig. 12.5). Such a drug has been found and has passed clinical tests.

Another less common approach makes use of redundancy in certain DNA repair systems of mammalians. In particular, single-stranded breaks can be repaired by the systems based on homologous recombination and by direct repair of the breaks. The latter repair system involves an enzyme called *PARP*. It was found that a mouse with

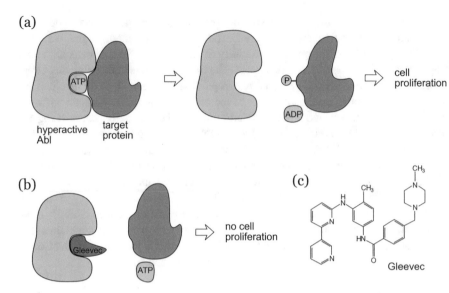

Fig. 12.5 Inhibition of an oncogenic protein by a small molecule. (**a**) The chimeric version of tyrosine kinase Abl, a key player in the development of chronic myelogenous leukemia, is hyperactive in the cancer cells. (**b**) The activity of Abl can be blocked by the drug Gleevec which binds the active center of the kinase. (**c**) Chemical structure of Gleevec

an inactivated PARP system develops normally because the other system makes the needed repair. So, normal cells do not suffer without the PARP system. It is very often that one or another DNA repair system is broken in cancer cells. In particular, the system that repairs single-stranded breaks by homologous recombination is often broken there, usually due to mutations in one of the key proteins called *Brca1* and *Brca2*. It was suggested that for such cells, inhibition of an alternative system, PARP, can have devastating consequences for the cancer cells, since they will not be able to repair single-stranded breaks. On the other hand, the inhibition would make no harm to normal cells. Therefore, inhibitors of PARP could serve as anticancer drugs for Brca-deficient cancers. This idea was confirmed experimentally. A few inhibitors of PARP have been developed. They have shown great efficiency in curing some Brca-deficient cancers, with relatively few side effects. At least three such inhibitors have been approved for the clinical treatment of ovarian cancer.

12.5.2 Immunotherapy of Cancer

It is a very attractive idea to use the immune system to fight cancer cells. As we describe in Chap. 10, the immune system of mammalians is extremely sophisticated in defense against infectious organisms. The system helps greatly in killing potential cancer cells that have not developed protection from this killing. But growing cancers somehow managed to neutralize the immune system. Treatment is needed to help the immune system in fighting cancer cells.

The simplest way to use immunotherapy against cancer is to find a cancer-specific marker on the surface of the cancer cell, develop antibodies to this marker, and inject them into the organism. Antibodies can have enormous specificity to their targets, which should limit the side effects of the therapy. The approach has been used successfully in some cases. It was found, for example, that a notable fraction of breast cancer cells express on their surface a very high level of Her2 receptor protein. Antibodies to Her2 have been developed successfully and are now in clinical practice. It is also possible to attach a poison to such antibodies, to enhance the efficiency of the therapy. Of course, the approach can be applied only if a selected marker on the cancer cell surface is rare on the surface of normal cells. This restricts the application of the approach.

A different approach attracts much attention now. As emphasized in Chap. 10, the major concern of the immune system is to avoid the self-destruction of the organism's cells. In particular, the T cells, which are involved in the first lines of the immune response, should not attack other cells of the organism. They have to bind and kill only "foreign" cells. The selectivity between own and foreign is achieved by trying to bind protein fragments exposed on the cell surface. Exposing at the cell surface fragments of proteins that are present in the cell is a common feature of all cells. T cells that can strongly bind fragments of the organism's proteins are selected to death at the very early stage of their development in the thymus. Cancer cells have, on average, around 50 mutated genes, which are by-products of cancer development. Due to these mutations, fragments of the corresponding proteins are foreign to the T cells. Since these fragments are exposed to the surface of the cancer cells, the cells could be recognized and killed by the T cells. It does not happen, however, because cancer cells are capable of restricting the response of T cells. The activity of T cells is regulated by a complex system of receptors on their surface which, when activated, are capable of inhibiting or stimulating their activity. This regulation is needed, in particular, as a backup system to prevent the T cell attack on the organism's normal cells. The normal cells expose special proteins on their surface which bind and activate the inhibitory receptors on the surface of T cells. The activated inhibitory receptors strongly restrict the T cell's ability to attack other cells. The same proteins, in an even larger amount, are exposed on the surface of cancer cells (Fig. 12.6). As a result, the inhibitory receptors on the surface of T cells bind the proteins activating them, preventing the attack of T cells on the cancer cells.

This picture suggests the following way to eliminate cancer cells. One can create special antibodies that attach to the inhibitory receptors to prevent their interaction with the proteins which activate them. So, the activity of the inhibitory receptors remains very low, allowing the T cells to kill the cancer cells. Since various protein fragments on the surface of cancer cells can be recognized by the T cells as foreign, cancer cells cannot avoid the T cell attack through an additional mutation of a single mutant protein. This makes it difficult for the cancer cells to avoid the attack of the T cells.

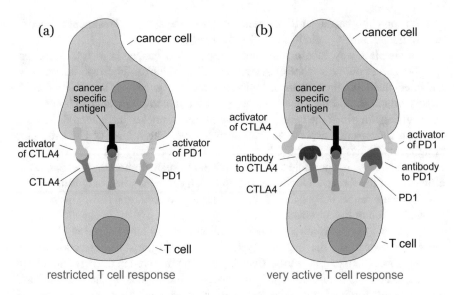

Fig. 12.6 Blocking the activation of inhibitory receptors on the surface of T cells. (**a**) The cancer cell exposes a cancer-specific antigen on its surface and, therefore, can be recognized and killed by a T cell. However, the response of the T cell is strongly restricted by the activation of its inhibitory receptors. The important inhibitory receptors, CTLA4 and PD1, are located on the surface of the T cell and activated by specific proteins exposed on the surface of cancer cells. (**b**) The activation of the inhibitory receptors can be prevented by special antibodies that block the interaction of the receptors with their activators. This allows T cells to attack and kill cancer cells

The approach described above is based on suppressing inhibitors of T cell activity. A complementary strategy, based on increasing the activity of T cell stimulators, is also under development. Other approaches to killing cancer cells by recruiting the immune system are also being tested. In general, the immunotherapy of cancer is rapidly gaining traction.

12.5.3   Multidrug Therapy

A key feature of cancer cells is their genetic instability. As we saw above, this instability can itself be a target of anticancer treatment. However, due to genetic instability, cancers can develop resistance to various types of therapy. In many cases, the cancer cells become undetectable after a treatment, but a tiny fraction of them somehow survived. These surviving cells may have resistance against the treatment because of the heterogeneity of the cancer cells, or maybe they were hidden in a protective environment. In the latter case, they may develop, due to accumulating mutations, a resistance to the initial therapy. This takes time, months, or even years, but eventually, these cells give rise to the same cancer, which is now resistant to the initial treatment. Such delayed resistance represents a huge problem for therapists fighting cancers. One solution here is the application of a few different treatments simultaneously. The probability that there are cells in cancer that are resistant to two or three different treatments is much lower than to any single treatment. Thus, a combination of therapies has a chance to kill all cancer cells. In some cases, this strategy allows the complete elimination of metastasizing cancers.

Glossary

acetyl CoA A small molecule that has a high-energy bond. When the bond is hydrolyzed, a large amount of free energy is released.

acid A compound that releases protons (H^+) and lowers the pH, when it is dissolved in water.

action potential Self-propagating electrical *excitation* i*n* a neuron or muscle cell. Action potentials make possible fast long-distance signaling in the nervous system.

activation energy The energy which must be acquired by atoms or molecules to undergo a particular chemical reaction (Fig. 2.2).

ADP (adenosine 5′-diphosphate) Nucleotide produced by hydrolysis of the terminal phosphate of ATP. Hydrolysis of ATP releases a large amount of free energy.

allele One of two alternative forms of a gene in a diploid cell. Each of the two homologous chromosomes has one allele of a gene.

allosteric transition Change in a protein's conformation caused by the binding of a regulatory ligand, or by covalent modification. The conformational change can involve the entire protein and alter its activity (Fig. 2.4).

alpha helix (α-helix) The common structural motif in proteins, in which a linear segment of amino acids folds into a right-handed helix stabilized by hydrogen bonds between atoms of the polypeptide backbone (Fig. 1.13).

amino acid A structural unit of proteins. Each molecule contains an amino group, a carboxyl group, and a side group linked to the same carbon atom (Fig. 1.12).

antibiotic A molecular that is toxic to microorganisms. It can be a natural product of a particular microorganism or plant, or an artificial substance.

antibody Protein produced by activated B cells in response to infection by a pathogen or foreign molecule. It binds tightly and very specifically to the pathogen or foreign molecule, inactivating it or marking it for destruction (Fig. 10.2).

anticodon Sequence of three nucleotides in a transfer RNA (tRNA) molecule that is complementary to three nucleotides of the codon in a messenger RNA (mRNA) molecule.

© The Editor(s) (if applicable) and The Author(s), under exclusive license to Springer Nature Switzerland AG 2023
A. Vologodskii, *The Basics of Molecular Biology*,
https://doi.org/10.1007/978-3-031-19404-7

antigen A molecule that can induce an adaptive immune response or that can bind to an antibody or T cell receptor.

apoptosis Programmed cell death mediated by special intracellular enzymes.

ATP (adenosine 5′-triphosphate) Nucleotide composed of adenine, ribose, and three phosphate groups. ATP is the principal carrier of energy in cells. Hydrolysis of the terminal phosphate group (formation of ADP) releases a large amount of free energy (Fig. 2.7).

axon A long cable-like part of the nerve cell that can rapidly conduct nerve impulses over long distances to deliver signals to other cells (Fig. 7.6).

B cell receptor The transmembrane protein on the surface of a B cell that serves as an antigen receptor.

beta sheet (β-sheet) Structural motif in proteins in which sections of the polypeptide chain form a nearly flat sheet, stabilized by hydrogen bonds between atoms of the polypeptide backbone (Fig. 1.13).

binding site Region on the surface of one macromolecule (usually a protein or nucleic acid) that can bind another molecule through noncovalent bonding.

catalyst A substance that can increase the reaction rate by lowing its activation energy.

cell cycle A chain of steps in cell development that ends with cell division. Chromosomes and other cell contents are duplicated during the cycle.

chromatin The complex of DNA, histones, and other proteins located in the nucleus of a eukaryotic cell.

chromosome A complex of a very long DNA molecule and many specific proteins that carry the hereditary information of an organism.

codon A segment of three nucleotides in a DNA or mRNA molecule that codes a specific amino acid for its incorporation into a growing protein chain.

CRISPR Defense mechanism in bacteria based on special small RNA molecules which mark invading viral genomes for destruction through complementary base pairing.

cyclic AMP (cAMP) The nucleotide that is generated from ATP by a special enzyme in response to various extracellular signals (Fig. 7.5).

cytoplasm Contents of a cell within its plasma membrane.

cytosol The fluid contained in the cell cytoplasm

deoxyribonucleic acid (DNA) Polynucleotide formed from covalently linked deoxyribonucleotides. The storage and carrier of hereditary information.

differentiation The process by which a cell is converted into a cell of a specialized type during its development.

diffusion Random movement of molecules through space due to thermal motion.

DNA ligase The enzyme that joins the ends of DNA strands by forming a covalent bond.

DNA methylation Covalent attachment of methyl groups to DNA.

DNA polymerase The enzyme that synthesizes DNA strands by extending its 3′-end. The enzyme uses a single-stranded DNA template as a guide.

DNA repair Processes of repairing the various accidental lesions that occur in DNA.

DNA replication Process of making a copy of DNA molecule.

DNA topoisomerases Enzymes that catalyze the passage of DNA segments through a temporal single-stranded or double-stranded break in another DNA segment. The break is resealed after the passage. The reaction allows for resolving all possible tangles which appear during DNA functioning (Fig. 2.6).

embryonic stem cells The cells derived from the early mammalian embryo. They are capable of differentiating to all specific cells of the body.

enzyme Protein that catalyzes a specific chemical reaction.

epigenetic inheritance The inheritance of phenotypic changes in a cell that do not result from changes in the nucleotide sequence of DNA. It is usually due to heritable modifications in chromatin such as DNA methylation and histone modifications.

eukaryote Organism composed of one or more cells that have a nucleus.

exon A coding segment of a eukaryotic gene that will be represented in mRNA or a final transfer, ribosomal, or other RNA molecules. An exon is usually surrounded by introns, noncoding DNA segments.

fat Specific energy-storage molecules in cells.

gamete A haploid male or female germ cell that can unite with another gamete of the opposite sex to form a diploid zygote.

gene A sequence of nucleotides in DNA that encodes either a single protein or RNA.

genetic code The correspondence between nucleotide triplets (codons) in DNA or RNA and amino acids in proteins.

genome DNA that carries all genetic information belonging to a cell or an organism.

genotype Full set of genetic information in an individual cell or organism.

germ cell A reproductive cell of a multicellular organism. The line of germ cells includes both the haploid gametes and their diploid precursors.

G-protein A trimeric protein that transmits a signal from its specific G-protein-coupled receptor to the next signal transmitters (Fig. 7.4).

G-protein-coupled receptor A cell surface receptor that is activated by its specific extracellular ligand or by light. It activates a G protein, starting a chain of signal transmissions inside the cell (Fig. 7.3).

GTP A source of free energy in the cell. Nucleotide releases a large amount of free energy on hydrolysis of its terminal phosphate group. Has an important role in protein synthesis and cell signaling. .

helper T cell Type of T cell that produces a signal for activating B cells and some other cells of the immune system.

histone One of a few small abundant proteins that form the nucleosome cores. Nucleosomes, where the DNA segment is wrapped around the core, constitute the first level of structural organization of eukaryotic chromosomes.

homologous Genes and proteins that are similar due to a common origin.

horizontal gene transfer Gene transfer between bacteria by occasionally released DNA segments. Although the transfer is rare, its role in evolution is enormous.

hydrogen bond Noncovalent bond in which a hydrogen atom is partially shared by two electronegative atoms, oxygen or nitrogen, in living cells.

hydrolysis A chemical reaction in which a molecule of water breaks a chemical bond.

hydrophilic A substance that is easily dissolved in water.

intron Noncoding region of a eukaryotic gene. It is transcribed into an RNA molecule during transcription and is later excised by RNA splicing.

ion channel Transmembrane protein complex that forms a channel across the cell membrane through which specific inorganic ions can diffuse.

lagging strand One of the two newly synthesized strands during DNA replication. DNA polymerase synthesizes the lagging strand as discontinuous segments that are later joined covalently by DNA ligase (Fig. 6.3).

leading strand One of the two newly synthesized DNA strands. DNA polymerase synthesizes the leading strand by continuous synthesis (Fig. 6.3).

ligand Any molecule that binds to a specific protein site or a site on another molecule.

lineage Genetic line of descent of an animal or plant.

lipid bilayer Double layer of lipid molecules that forms the core of the cell membrane (Fig. 3.1).

lymphocyte White blood cell which is a major player in the adaptive immune system. Two main types of lymphocytes are B cells and T cells. B cells produce antibodies. T cells kill cells infected by viruses and bacteria and activate B cells.

macromolecule A molecule containing a very large number of atoms. Nucleic acids and proteins are examples.

macrophage A cell of the innate immune system. It engulfs and digests pathogens and activates helper T cells of the adaptive immune system.

malignant Tending to infiltrate and metastasize (about tumor cells).

master transcription regulator A transcription regulator which interacts with many genes and defines the cell type.

membrane potential Is the difference in electric potential between the interior and the exterior of a membrane. It is due to a slight excess of positive ions over negative on the membrane exterior and negative ions over positive on the membrane interior.

memory cell A long-living T or B lymphocyte appearing after antigen stimulation. It provides a faster response of the adaptive immune system on a later attack by the same antigen.

messenger RNA (mRNA) RNA molecule that specifies the sequence of amino acids in a protein.

metastasis Spread of cancer cells from the initial site of the tumor to other sites in the body.

MHC complex (major histocompatibility complex) A set of genes that code the transmembrane proteins involved in antigen presentation to T cells.

microRNAs (miRNAs) Eukaryotic RNAs (~21 nucleotides) that prevent the production of a particular protein by base pairing with mRNA that codes the protein and marks this mRNA for distraction.

microtubules Microscopic tubular structures that form an extended network in eukaryotic cells.

morphogenesis Processes by which shape and organs are created in the developing organism.

mutation Heritable change in the nucleotide sequence of DNA.

mutation rate The rate at which changes occur in DNA sequences.

myosin A motor protein that uses the energy of ATP hydrolysis to move along actin filaments, a major player in muscle contraction.

natural killer cell (NK cell) A cell of the innate immune system that can kill virus-infected cells and some cancer cells.

negative feedback Mechanism of regulation whereby the end product of a reaction inhibits its production.

neuron (nerve cell) Impulse-conducting cell of the nervous system (Fig. 7.6).

noncoding RNA An RNA molecule that does not code for a protein.

nuclease A protein of a large family of enzymes that cut DNA chains.

oncogene A gene in which a mutation can make the product of the gene stimulating cancer development.

operator DNA segment which contains the code necessary to begin transcription of the gene. It also can bind the gene repressor preventing its expression.

operon A cluster of genes that are transcribed together to give a single mRNA molecule.

p53 A tumor suppression protein that plays a key role in controlling cell division and cell death.

pathogen An organism, cell, or virus that causes disease.

phagocytosis Engulfing a large particle by the cell's plasma membrane and forming an internal compartment called the phagosome (Fig. 10.7).

phagosome A large intracellular compartment with a pathogen surrounded by the plasma membrane (Fig. 10.7).

phenotype The set of observable characteristics of a cell or organism.

piRNAs Small noncoding RNAs produced in the germ cells that prevent the movement of transposable elements by silencing their genes.

pluripotent stem cell A cell that can develop into nearly all types of cells or tissues in the adult body.

polypeptide Linear polymer consisting of amino acids.

positive feedback Mechanism of regulation whereby the end product of a reaction stimulates its production.

primary structure The linear sequence of monomer units in a polymer, such as the sequence of nucleotides in DNA.

programmed cell death A form of cell death in which a cell kills itself by activating an intracellular death program.

prokaryote Single-cell microorganism that lacks a membrane-enclosed nucleus. Both bacteria and archaea are prokaryotes.

promoter DNA segment to which RNA polymerase binds to begin transcription.

protein A linear chain of amino acids which are linked together in a specific sequence. Usually, the chain folds into a specific 3D structure.

purifying selection Natural selection which constantly eliminates individuals carrying deleterious mutations in their DNA.

reading frame The phase of reading nucleotides in sets of three. Only one of three possible phases encodes a required protein (Fig. 1.16).

receptor Protein that binds a specific signal molecule and initiates a response in the cell.

red blood cell A cell containing molecules of hemoglobin, a protein that carries oxygen from the lungs to all tissues of the body.

replication origin A segment of DNA molecule at which replication complex is assembled and the replication begins.

restriction nuclease A protein of the large family of nucleases that can cleave a DNA molecule at a specific short sequence of nucleotides.

reverse transcriptase The enzyme that catalyzes the formation of a double-strand DNA on a single-strand RNA template.

ribonucleic acid (RNA) Polynucleotide formed from covalently linked ribonucleotides. RNA molecules perform multiple functions in protein synthesis (Fig. 1.10).

ribosome A large particle composed of RNAs and ribosomal proteins that catalyzes the synthesis of proteins on mRNA templates.

RNA editing RNA processing that alters the nucleotide sequence of an RNA molecule after it is synthesized by inserting, deleting, or changing individual nucleotides.

RNA interference (RNAi) A phenomenon where small RNA molecules block protein translation. They achieve it by binding to the complementary segments of messenger RNAs that code for those proteins.

RNA polymerase The enzyme that catalyzes the synthesis of an RNA molecule on a DNA template.

RNA primer A short stretch of RNA synthesized on a DNA template. During replication, the DNA polymerase starts DNA synthesis from these primers.

RNA splicing Process in which intron sequences are excised from RNA transcripts and exons are joined together. The process leads to the formation of messenger RNAs (mRNAs).

secondary structure A regular local folding pattern of a polymer chain. Both nucleic acids and proteins can form sequence-dependent secondary structures.

side chain The part of an amino acid that differs one amino acid from another. The side chains define the unique physical and chemical properties of each amino acid.

sister chromatids Tightly linked pairs of chromosomes obtained by chromosome duplication in germ cells.

site-specific recombination A type of DNA recombination that requires a specific enzyme and short, specific sequences of DNA (Fig. 4.5).

small interfering RNAs (siRNAs) Short double-stranded RNAs that block protein translation by binding with mRNA coding the protein.

somatic cell Any cell of an organism other than cells of the germ line.

stem cell An undifferentiated cell that can continue dividing indefinitely. A daughter cell can either differentiate or remain a stem cell (Fig. 8.13).

substrate Molecule on which an enzyme acts.

telomerase The enzyme that elongates telomere sequences in DNA at the ends of eukaryotic chromosomes.

telomere A DNA segment with a short sequence repeat at the ends of a eukaryotic chromosome. It is replicated by the special enzyme, telomerase.

template A single strand of DNA or RNA which is used as a guide for the synthesis of a complementary strand.

tertiary structure 3D structure of a folded polymer chain. The majority of proteins and many RNA molecules have well-defined 3D structures.

transcription Synthesizing RNA molecule on a DNA strand by the enzyme RNA polymerase.

transcriptional control Regulation of gene expression by controlling its transcription.

transfer RNAs Small RNA molecules serving as an adaptor between mRNA and amino acids in protein synthesis (Fig. 1.18).

translation Process of mRNA-directed protein synthesis on a ribosome.

transmembrane protein A protein that extends through the cell membrane and has parts on both sides of the membrane.

transporter Membrane transport protein that binds to a specific molecule and transports it across the membrane in a specific direction by using the free energy of ATP hydrolysis coupled with the transport.

tumor suppressor gene A gene that assists in cancer prevention.

virus A particle consisting of nucleic acid (RNA or DNA) enclosed in a protein shell. Viruses cannot replicate themselves and have to use a host cell apparatus for their replication.

X chromosome The chromosome that specifies the distinctive features of female mammals.

zygote Diploid cell obtained by fusion of male and female gametes.

Index

n the United States
& Taylor Publisher Services

Printed
by Bake